BIRKHÄUSER

Modeling and Simulation in Science, Engineering and Technology

Series Editor
Nicola Bellomo
Politecnico di Torino
Italy

Advisory Editorial Board

M. Avellaneda (Modeling in Economics)
Courant Institute of Mathematical Sciences
New York University
251 Mercer Street
New York, NY 10012, USA
avellaneda@cims.nyu.edu

K.J. Bathe (Solid Mechanics)
Department of Mechanical Engineering
Massachusetts Institute of Technology
Cambridge, MA 02139, USA
kjb@mit.edu

P. Degond (Semiconductor and Transport
  Modeling)
Mathématiques pour l'Industrie et la Physique
Université P. Sabatier Toulouse 3
118 Route de Narbonne
31062 Toulouse Cedex, France
degond@mip.ups-tlse.fr

A. Deutsch (Complex Systems
  in the Life Sciences)
Center for Information Services
  and High Performance Computing
Technische Universität Dresden
01062 Dresden, Germany
andreas.deutsch@tu-dresden.de

M.A. Herrero Garcia (Mathematical Methods)
Departamento de Matematica Aplicada
Universidad Complutense de Madrid
Avenida Complutense s/n
28040 Madrid, Spain
herrero@sunma4.mat.ucm.es

W. Kliemann (Stochastic Modeling)
Department of Mathematics
Iowa State University
400 Carver Hall
Ames, IA 50011, USA
kliemann@iastate.edu

H.G. Othmer (Mathematical Biology)
Department of Mathematics
University of Minnesota
270A Vincent Hall
Minneapolis, MN 55455, USA
othmer@math.umn.edu

L. Preziosi (Industrial Mathematics)
Dipartimento di Matematica
Politecnico di Torino
Corso Duca degli Abruzzi 24
10129 Torino, Italy
luigi.preziosi@polito.it

V. Protopopescu (Competitive Systems,
  Epidemiology)
CSMD
Oak Ridge National Laboratory
Oak Ridge, TN 37831-6363, USA
vvp@epmnas.epm.ornl.gov

K.R. Rajagopal (Multiphase Flows)
Department of Mechanical Engineering
Texas A&M University
College Station, TX 77843, USA
krajagopal@mengr.tamu.edu

Y. Sone (Fluid Dynamics in Engineering
  Sciences)
Professor Emeritus
Kyoto University
230-133 Iwakura-Nagatani-cho
Sakyo-ku Kyoto 606-0026, Japan
sone@yoshio.mbox.media.kyoto-u.ac.jp

Antonio Romano

Geometric Optics

Theory and Design of Astronomical Optical Systems Using Mathematica®

Birkhäuser
Boston • Basel • Berlin

Antonio Romano
Dipartimento di Matematica
e Applicazioni "R. Caccioppoli"
Università degli Studi di Napoli "Federico II"
via Cintia
80126 Napoli
Italy
antroman@unina.it

ISBN 978-0-8176-4871-8 e-ISBN 978-0-8176-4872-5
DOI 10.1007/978-0-8176-4872-5
Springer New York Dordrecht Heidelberg London

Library of Congress Control Number: 2009938171

Mathematics Subject Classification (2000): 78A05, 78-01, 78-04

© Birkhäuser Boston, a part of Springer Science+Business Media, LLC 2010
All rights reserved. This work may not be translated or copied in whole or in part without the written permission of the publisher (Birkhäuser Boston, c/o Springer Science+Business Media LLC, 233 Spring Street, New York, NY 10013, USA), except for brief excerpts in connection with reviews or scholarly analysis. Use in connection with any form of information storage and retrieval, electronic adaptation, computer software, or by similar or dissimilar methodology now known or hereafter developed is forbidden.
The use in this publication of trade names, trademarks, service marks and similar terms, even if they are not identified as such, is not to be taken as an expression of opinion as to whether or not they are subject to proprietary rights.

Printed on acid-free paper

Birkhäuser Boston is part of Springer Science+Business Media (www.birkhauser.com)

Contents

Preface ix

1 Fermat's Principle and General Considerations Regarding Centered Optical Systems 1
   1.1 An Introduction to Fermat's Principle 1
   1.2 Forming an Image 5
   1.3 Axial Chromatic Aberrations 9
   1.4 Monochromatic Aberrations 10
   1.5 Surfaces of Revolution 13
   1.6 Ray Tracing in Axially Symmetric Systems 17
   1.7 Exercises 20

2 Gaussian Optics 25
   2.1 Gaussian Approximation for a Single Surface 25
   2.2 Compound Systems 29
   2.3 Principal Planes and Focal Lengths 32
   2.4 Stops and Pupils 34
   2.5 Some Gaussian Optical Invariants 36
   2.6 Gaussian Analysis of Compound Systems 38
   2.7 A Graphical Method 42
   2.8 Exercises 43

3 Fermat's Principle and Third-Order Aberrations 49
   3.1 Introduction 49
   3.2 The Aberration Function 50
   3.3 A New Aberration Function 54
   3.4 The Aberration Function Φ for a Single Surface 59
   3.5 The Total Aberration Function for a Compound System . . 63
   3.6 Analysis of Third-Order Aberrations 67
   3.7 Petzval's Theorem 72
   3.8 Aberration Formulae 74

3.9	Combined Effect of Third-Order Aberrations	75
3.10	Primary or Axial Chromatic Aberrations	76
3.11	Aplanatism and the Helmholtz Condition	79
3.12	Some Applications	83
3.13	Light Diffraction and the Airy Disk	85

4 Newtonian and Cassegrain Telescopes — 91

4.1	Newtonian Telescopes	91
4.2	Cassegrain Telescopes	94
4.3	Spherical Aberration and Coma in Cassegrain Telescopes	96
4.4	Concluding Remarks	98
4.5	Examples	98

5 Cameras for Astronomy — 101

5.1	Introduction	101
5.2	Aberrations for a Single Mirror	102
5.3	Schmidt Cameras	103
5.4	Examples	107
5.5	Wright Cameras	109
5.6	Houghton Cameras	110
5.7	Examples	120
5.8	Maksutov Cameras	121
5.9	Examples	124

6 Compound Cassegrain Telescopes — 127

6.1	Introduction to Cassegrain Telescopes	127
6.2	Schmidt–Cassegrain Telescopes	129
6.3	Examples	131
6.4	Houghton–Cassegrain Telescopes	133
6.5	Examples	137
6.6	Maksutov–Cassegrain Telescopes	140
6.7	Examples	142
6.8	Final Remarks	144

7 Doublets and Triplets — 145

7.1	Achromatic Doublets	145
7.2	Elimination of Spherical Aberration and Coma	147
7.3	Examples	149
7.4	Triplets	152
7.5	Examples	154

8 Other Optical Combinations — 157
- 8.1 Cassegrain Telescope with Spherical Surfaces 157
- 8.2 The Flat-Field Baker–Schmidt Camera 159
- 8.3 Cassegrain Telescope with the Corrector at the Prime Focus 162
- 8.4 The Klevtsov Telescope 164
- 8.5 A First Analysis of the Meniscus 165
- 8.6 Analysis of the Mangin Mirror 167
- 8.7 Buchroeder Camera 168

9 Fermat's Principle and Wavefronts — 171
- 9.1 Fermat's Principle 171
- 9.2 The Boundary Value Problem 180
- 9.3 Rotational Symmetry and Lagrange's Invariant 183
- 9.4 Wavefronts and Fermat's Principle 185
- 9.5 Huygens' Principle 188

10 Hamiltonian Optics — 191
- 10.1 Hamilton's Equations in Geometrical Optics 191
- 10.2 Hamilton's Principal Functions 192
- 10.3 Symmetries and Characteristic Functions 197
- 10.4 Lagrange's Optical Invariant for Axially Symmetric Systems 199

11 Monochromatic Third-Order Aberrations — 201
- 11.1 Introduction to Third-Order Aberrations 201
- 11.2 Third-Order Aberrations via the Angular Function 202
- 11.3 Reduced Coordinates 206
- 11.4 Schwarzschild's Eikonal 207
- 11.5 Addition Theorem for Third-Order Aberrations 210
- 11.6 Fourth-Order Expansion of the Angular Function 212
- 11.7 Aberrations of Axially Symmetric Systems 215

References — 219

Index — 223

Preface

A very wide selection of excellent books are available to the reader interested in geometric optics. Roughly speaking, these texts can be divided into three main classes.

In the first class (see, for instance, [1]–[9]), we find books that present the theoretical aspects of the subject, usually starting from the Lagrangian and Hamiltonian formulations of geometric optics. These texts analyze the relations between geometric optics, mechanics, partial differential equations, and the wave theory of optics. The second class comprises books that focus on the applications of this theory to optical instruments. In these books, some essential formulae (which are reported without providing proofs) are used to propose exact or approximate solutions to real-world problems (an excellent example of this class is represented by [10]). The third class contains books that approach the subject in a manner that is intermediate between the first two classes (see, for instance, [11]–[16]).

The aim of this book, which could be placed in the third class, is *to provide the reader with the mathematical background needed to design many optical combinations that are used in astronomical telescopes and cameras.*[1] The results presented here were obtained by using a different approach to third-order aberration theory as well as the extensive use of the software package ***Mathematica**®*.

The different approach to third-order aberration theory adopted in this book is based on Fermat's principle and on the use of particular optical paths (not rays) termed *stigmatic paths*. This approach makes it easy to derive the third-order aberration formulae. In this way, the reader is able to understand and handle the formulae required to design optical combinations without resorting to the much more complex Hamiltonian formalism and Seidel's relations. On the other hand, the Hamiltonian formalism has unquestionable theoretical utility considering its important applications in

---

[1] For a good example of a professional textbook on astronomical optics, see [13].

optics, in mechanics, and in the theory of partial differential equations. For this reason, Hamiltonian optics is widely discussed in Chapters 9–11.

The use of **Mathematica**® to design optical combinations is shown to be very convenient. In fact, although the aberration formulae are obtained in an elementary way, their application in the design process necessitates a lot of calculations. Using **Mathematica**®, it is possible to implement programs that allow us to realize the third-order design of all the astronomical combinations described in this book. Although experience has shown that a design based on third-order optics is not always acceptable, this approach can be used as a starting point for any optimization method available in professional software, such as **OSLO** and **ATMOS**, the simplest versions of which can be freely downloaded from the Internet. However, we must bear in mind that optimization methods will only give correct results if the data used in the approximate design are very similar to those used in the final project. These methods must be handled with great care, since they will very often lead to a new design that is worse than the original one. The reason for this is that the function to be minimized contains many minima that are very close to each other and do not correspond to an effective improvement in the optical combination. For this reason, the author, with the help of A. Limongiello, developed the software **Optisoft** (which runs in the Microsoft Windows environment), which allows the *final* forms of all the optical combinations considered in this book to be obtained.

In the first chapter, the essential aspects of an optical system $\mathbb{S}$ with an axis of rotational symmetry are introduced. Moreover, we analyze all of the data supplied by optical software in order to *check* whether a given optical system $\mathbb{S}$ is acceptable or not. Chapter 2 describes the Gaussian characteristics of $\mathbb{S}$: conjugate planes, magnification, focal and nodal points, principal planes and optical invariants. The matrix form of the Gaussian approximation is also presented in detail. All of the Gaussian data for an optical system can be derived using the notebook **GaussianData**.[2]

In Chapter 3, a new approach to third-order monochromatic aberration that is based on both Fermat's principle and *stigmatic paths* is described. Here it is shown that these optical paths can be used in Fermat's principle instead of the real rays, with the advantage that the stigmatic paths are completely known, since they are determined by Gaussian optics. The third-order aberrations for any optical system can be obtained in mathematical form using the notebook **TotalAberrations**. It should be noted that the symbolic formulae are so dense for optical systems containing many elements with finite thicknesses that they are not practical to apply.

---

[2] A program written with **Mathematica**® can be saved as a notebook or a package.

Chapter 4 contains an analysis of Newtonian and Cassegrain telescopes based on conical mirrors. In Chapters 5 and 6 we study photographic cameras containing lenses and mirrors (Schmidt, Wright, Houghton, and Maksutov cameras), as well as the corresponding catadioptric Cassegrain telescopes. Finally, the third-order design of achromatic doublets or apochromatic doublets and triplets is discussed in Chapter 7. Some other interesting optical devices, including the Klevtsov combination and the Baker–Schmidt flat-field camera, are studied in Chapter 8.

Finally, the Lagrangian and Hamiltonian formulations for geometric optics and Seidel's third-order aberration theory are treated in Chapters 9–11.

Each optical combination analyzed in this book is accompanied by a notebook that automates its third-order design. All of these notebooks work in versions 4, 5, and 6 of ***Mathematica*®** and may be downloaded from the Publisher's website at: http://www.birkhauser.com/978-0-8176-4871-8. These notebooks represent an integral part of the book for many reasons. First, they contain many calculations that appear in the book and many worked exercises. Moreover, many other exercises can be carried out by the reader him- or herself. Finally, carefully studying the programs contained in the notebooks could provide a useful way for readers to learn how to program with ***Mathematica*®**.

We conclude by noting that amateurs with sufficient knowledge of mathematics may find it interesting to learn how to derive the formulae listed in many manuals from the general laws of geometric optics. On the other hand, amateurs who are not interested in learning the mathematical background of optics can use the notebooks contained in the book to rapidly obtain the third-order designs of many cameras and telescopes used in astronomy.

Naples, Italy *Antonio Romano*
August 2009

Chapter 1

Fermat's Principle and General Considerations Regarding Centered Optical Systems

1.1 An Introduction to Fermat's Principle

Faraday and Lentz discovered that a varying magnetic field produces an electric field. Subsequently, Maxwell hypothesized that the inverse process was possible—that a varying electric field will produce a magnetic field—and he postulated a set of equations relating to electromagnetism in his famous *Treatise on Electricity and Magnetism*. Ever since then, electric and magnetic phenomena have been regarded as two interconnected aspects of the electromagnetic field, which, in turn, is propagated by electromagnetic waves. Moreover, Maxwell showed that light waves are a particular type of electromagnetic wave, so the field of optics merged with electromagnetism. This suggests that one could study the propagation of light waves across an optical system using Maxwell's equations along with suitable boundary conditions, depending on the nature of the media involved. However, although theoretically possible, such an approach is not practical, since it introduces insurmountable technical difficulties. For example, the simple analysis of light propagation across a hole in an obstacle (i.e., of the diffraction of light) is so complex that it is necessary to apply the approximate procedure expressed by Huygens–Kirchhoff's principle (see Section 9.5).

Consequently, we must follow an approach that is frequently used in physics—we must resort to using a simpler model to describe such phenomena. **Geometric optics**, which amounts to a drastic simplification of light propagation, is applicable when the dimensions of the bodies across which the light propagates are large compared with the wavelength of the light.

Let us consider the propagation of a monochromatic light wave in an isotropic medium Σ that has a refractive index $N(\mathbf{x})$ which depends on the point **x**. If we suppose that the wavelength of the light is small compared with the dimensions of both the interposed obstacles ("stops") and the regions traversed, the propagation of the light can be described by either **wavefronts** (**Huygens' principle**) or by **rays**, which are the curves normal to wavefronts (**Fermat's principle**).

For now, we will postulate Fermat's principle without criticism and apply it in order to derive Gaussian optics and third-order aberration theory in an elementary way. The results of the first two chapters will then be used to analyze and design many astronomical optical systems. Finally, the theoretical aspects of geometric optics will be developed in Chapters 9–11 of this book.

The **optical path length** $OPL(\gamma)$ along *any curve* γ is defined by the following formula:

$$OPL(\gamma) = \int_\gamma N(\mathbf{x})ds. \qquad (1.1)$$

If the curve γ is a light ray, v denotes the speed of light in medium Σ, and c is the speed of light in a vacuum, then we have

$$Nds = \frac{c}{v}ds = cdt,$$

so that the elementary optical path length along a *ray* is proportional to the time dt taken for the light to propagate along γ.

The starting point for geometric optics is **Fermat's principle**, which (very roughly) can be formulated in the following way:

The optical path length of a light ray passing from point* x *to point* x' *is the length of the shortest optical path between the same points.

We must now answer the following fundamental questions:

- Does this principle determine the ray γ between two given points **x** and **x'** when the refractive index $N(\mathbf{x})$ has been assigned?

- Is the ray uniquely determined?

- How do we take into account the discontinuities in the refractive index $N(\mathbf{x})$ along the path of the ray?

Fermat's principle of the minimum optical path length was one of the first examples of a variational principle in physics. Since it was formulated, the number of variational principles has grown substantially in both physics and mathematics, leading to problems that do not (for the most part) currently have explicit exact solutions.

Briefly, a variational principle characterizes the unknown of a problem (for instance, the ray between two points **x** and **x'**) by requiring that it minimizes a suitable integral expression (for instance, the optical path length).

1.1. An Introduction to Fermat's Principle

This kind of principle leads to a mathematical problem that is quite different in nature from the initial value problem that Cauchy stated for ordinary differential equations. In the latter, we search for a function that satisfies suitable initial data and, *for any value of the independent variable*, a differential equation (i.e., a relation between the function itself and its derivatives). However, it is not appropriate here to dive deeply into the subject of variational principles, presenting their abstract and general formulations. The aim here is simply to remind the reader that many problems expressed in terms of differential equations could, equivalently, be formulated by resorting to a variational principle. Initially, the fact that nature seems to verify this "minimum principle" suggested to some that it had a divine origin, and so they attributed a metaphysical meaning to variational principles.

We answer all of the above questions in Chapter 9. In particular, it will be shown that a ray is characterized by the condition that the optical path length is *stationary* along it; this does not necessarily imply a minimum. However, for now, the contents of the first eight chapters are based on the following remarks, which derive from Fermat's principle.

1. ***In regions with a constant refractive index, rays are straight lines.***

   In fact, in these regions, the optical path length $OPL(\gamma)$ of a curve γ between two points $\mathbf{x}_1$ and $\mathbf{x}_2$ is proportional to the geometric length of γ. Consequently, the minimum value of $OPL(\gamma)$ corresponds to the length of the shortest curve between $\mathbf{x}_1$ and $\mathbf{x}_2$, which is a straight line.

2. ***Let S be a surface separating two homogeneous media with refractive indices N and N'. If $\mathbf{x}$ is the point at which a ray meets S, $\mathbf{t}$, $\mathbf{t}'$ denote the tangent unit vectors to the ray at $\mathbf{x}$ in the first and in the second media, respectively, and $\mathbf{n}$ is the internal unit normal to S at $\mathbf{x}$, then the vectors $\mathbf{t}$, $\mathbf{t}'$, and $\mathbf{n}$ verify the equation*** (see Exercise 1):

$$N'\mathbf{t}' - N\mathbf{t} = \lambda \mathbf{n}, \tag{1.2}$$

   where λ is a scalar quantity. In particular, the vectors $\mathbf{t}$, $\mathbf{t}'$, and $\mathbf{n}$ are coplanar. Moreover, after denoting the incident and refractive angles (i.e., the angles that $\mathbf{t}$ and $\mathbf{t}'$ form with $\mathbf{n}$; see Figure 1.1) by i and i', respectively, and evaluating the vector and the scalar products of (1.2) with $\mathbf{n}$, the following relations are obtained:

$$\sin i' = \frac{N}{N'} \sin i, \tag{1.3}$$

4   Chapter 1. Fermat's Principle and General Considerations

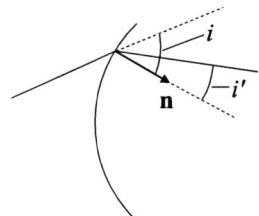

Fig. 1.1 The refraction law

$$\lambda = N'\sqrt{1 - \frac{N^2}{N'^2}\sin^2 i} - N\cos i. \qquad (1.4)$$

Relations (1.2)-(1.4), which express the **refraction law**, will be considered again in Chapter 9.

3. **Let S be a surface separating two regions C_1 and C_2 with constant refractive indices N_1 and N_2, respectively. We denote the family of all broken lines starting from the point $x_1 \in C_1$ and ending at the point $x_2 \in C_2$ by Γ. If the optical path length $OPL(\gamma)$, with $\gamma \in \Gamma$, is independent of γ, then all the broken lines of Γ are rays.**

In fact, since $OPL(\gamma)$ has the same value for any $\gamma \in \Gamma$, it is stationary along any path.

Under these conditions, the initial and final points of the rays belonging to Γ are said to be **stigmatic** with respect to S.

4. **Let $I(\mathbf{x})$ be a spatial region containing the point $\mathbf{x}$. Suppose that for any point $\mathbf{x}' \in I(\mathbf{x})$ one and only one ray γ exists between $\mathbf{x}$ and $\mathbf{x}'$. Then, denoting the unit tangent vector to γ by $\mathbf{t}$, the optical path length along γ**

$$V(\mathbf{x}, \mathbf{x}') = \int_{\mathbf{x}}^{\mathbf{x}'} N\mathbf{t} \cdot d\mathbf{x}, \qquad (1.5)$$

is a function of $\mathbf{x}$ and $\mathbf{x}'$. Moreover, the following relations hold:

$$\nabla_{\mathbf{x}} V(\mathbf{x}, \mathbf{x}') = -N(\mathbf{x})\mathbf{t}(\mathbf{x}), \quad \nabla_{\mathbf{x}'} V(\mathbf{x}, \mathbf{x}') = N(\mathbf{x}')\mathbf{t}(\mathbf{x}'). \qquad (1.6)$$

Formulae (1.6), which derive from Leibniz's rule, will be proven in Chapter 9.

Remark We note that the **reflection law**, according to which the angle of incidence i and the angle of reflection i' satisfy the condition $i' = -i$, can be obtained by the refraction law (1.3) by inserting $N = -N' = 1$ into it.

In Chapters 2–8 we show many applications of the above consequences of Fermat's principle.

1.2 Forming an Image

Throughout this book we consider a centered optical system $\mathbb{S}$ consisting of surfaces with a common axis a of revolution that separate homogeneous optical media with different (but constant for each medium) refractive indices. The axis a is called the **optical axis** of $\mathbb{S}$. We also suppose that the object that $\mathbb{S}$ must image is present in the plane π, which is orthogonal to a and is called the **object plane**.

All of the considerations discussed in the following sections of this chapter refer to a **given** optical system $\mathbb{S}$ with the above characteristics.

To analyze the behavior of the rays originating from a point on the object and crossing the optical system $\mathbb{S}$, we can use the refraction and reflection laws. This procedure, which is called **ray tracing**, will be analyzed in detail in Section 1.6. It is evident that it requires that the mathematical quantities that define a ray are assigned. Now, if we introduce Cartesian coordinates O, x, y in the object plane π, then a ray γ coming from (x, y) is determined by providing the unit vector $\mathbf{t}$ along γ, i.e., the two director cosines α and β of γ (see Figure 1.2).

It should be noted that the quantities $x, y, \mathbf{t}$, although natural, are not the most convenient variables to use. In fact, it is evident that some rays from $(x, y) \in \pi$ do not reach the optical system $\mathbb{S}$ due to either the finite dimensions of the optical components or the presence of a stop along the optical path. Let us consider the set of the rays arising from the point $O \in a$.

In Figure 1.3, the rim of the first surface of the optical path limits the aperture of the cone C of rays from O. It is then natural to identify a ray γ by providing the coordinates (x, y) of the point on the object in π and the coordinates (x_e, y_e) of the point of intersection of γ with the tangent plane π_e at the vertex of the first surface of $\mathbb{S}$. In Figure 1.4, the cone C is determined by a stop in the plane π_e, and we can again identify (x_e, y_e) as the coordinates of the point of intersection of a ray with π_e. However, in Figure 1.5 the stop is located after the optical system. In this case, owing to Fermat's principle, a ray from O is still determined by providing the

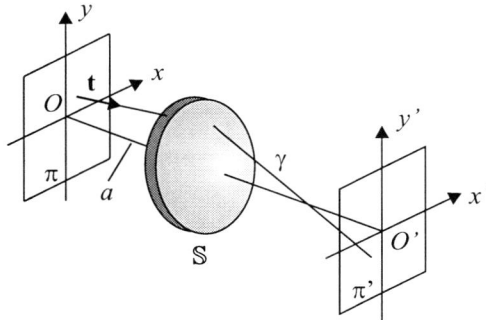

Fig. 1.2 A ray γ from the object plane to the image plane

coordinates of its point of intersection with the plane of the stop, but we cannot explicitly write the equation of the ray by providing these coordinates. In this case, we must choose a suitable plane π_e in front of the object plane π, and determine the coordinates (x_e, y_e) of the point of intersection of γ with π_e of any ray which *succeeds in crossing* $\mathbb{S}$. In Chapter 2 we discuss the criterion for defining a suitable reference plane π_e. This analysis, which leads us to the definitions of **entrance pupil** and **exit pupil** in particular, shows that the reference plane is just given by π_e in the cases of Figures 1.3 and 1.4, whereas it gives the correct position of the plane π_e in the case of Figure 1.5. In conclusion, from now on, a ray will be identified by the variables x, y, x_e, y_e.

Let π' be another plane orthogonal to the optical axis a. Two points **x** and **x'** are said to be **stigmatic for** $\mathbb{S}$ if all of the rays from **x** reach the point **x'** after crossing the system $\mathbb{S}$. The system $\mathbb{S}$ is said to be an **ideal system** or an **aberration-free system** for the object plane π if the following conditions are satisfied:

1. There is a plane π' such that points in π exhibit a one-to-one correspondence with points in π', independent of the wavelength of the light. In other words, each point $\mathbf{x} \in \pi$ corresponds to one and only one point $\mathbf{x}' \in \pi'$, and so the points **x** and **x'** are stigmatic for $\mathbb{S}$.

2. The above correspondence preserves the form of the object.

In this case, the plane π' is said to be the **image plane** and π and π' are said to be **conjugate planes**.

Consider two Cartesian frames $Oxyz$ and $Ox'y'z'$ that have their origins O, O' in π, π', respectively, and that have the axes Oz, Oz' coinciding with the optical axis a of $\mathbb{S}$. The coordinate plane Oyz is called the **meridional** or **tangential plane**, whereas Oxz is said to be the **sagittal plane**. Due

1.2. Forming an Image

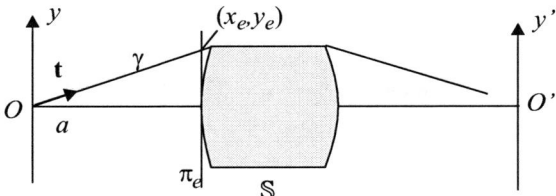

Fig. 1.3 Stop on the first surface

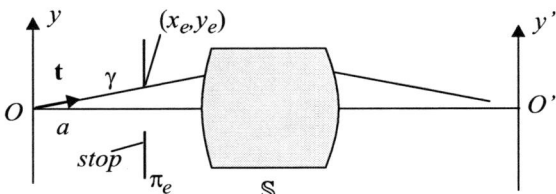

Fig. 1.4 Stop before the first surface

to the refraction law, the rays that are present in the plane Oyz before meeting $\mathbb{S}$ remain in this plane after crossing $\mathbb{S}$. They are called **meridional** or **tangential rays**, whereas the rays that lie in the plane Oxz are said to be **sagittal rays** (see Figures 1.6 and 1.7). All of the other rays are called **skew rays**.

The region containing the object, which by convention is always positioned on the left-hand side of $\mathbb{S}$, is the **object space**, whereas the region containing the image formed by $\mathbb{S}$ is the **image space**.

If π_e is the abovementioned auxiliary plane in the object space, then any ray r coming from the point (x, y) in the object plane π is completely determined by providing the Cartesian coordinates (x_e, y_e) or the polar coordinates (r, φ) of the point at which r intersects with π_e. Let (x', y') be the point in the image plane π' reached by r. Using this notation, we can say that the system $\mathbb{S}$ is ideal if it results in:

$$\begin{aligned} x' &= Mx, \\ y' &= My, \end{aligned} \quad (1.7)$$

where M, which is called the **magnification** of $\mathbb{S}$ with respect to the pair of conjugate planes π and π', is independent of (x, y), (x_e, y_e), and the wavelength of light λ (i.e., of the point on the object, the ray, and the wavelength).

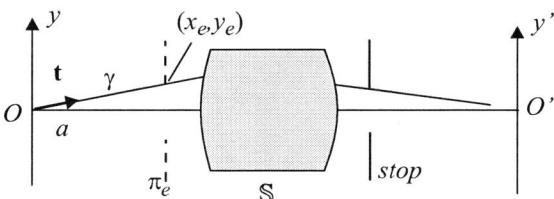

Fig. 1.5 Stop behind the last surface

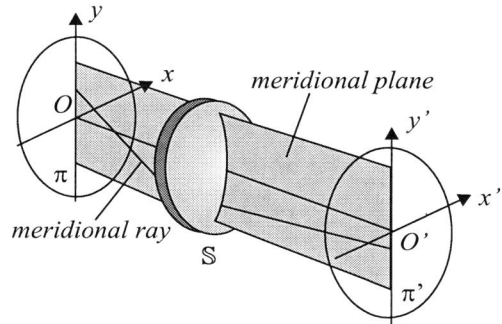

Fig. 1.6 Meridional rays

In other words, an optical system is ideal if the the following properties hold:

- The image of the object is planar and does not exhibit distortion

- Any point on the object and its corresponding point on the image are stigmatic

- The above properties are independent of the wavelength of light λ.

In the following chapter it is shown that, *for a given wavelength λ and object plane π*, any centered optical system is almost ideal; in other words, there is a unique image plane π'—termed the **Gaussian image plane**—for which the first two of the above properties are true, at least for points on the object and image belonging to two small neighborhoods of π and π' around the optical axis (the **Gaussian or paraxial approximation**). However, in this case, both the position of the ideal image plane π' and the magnification depend on λ.

1.3. Axial Chromatic Aberrations

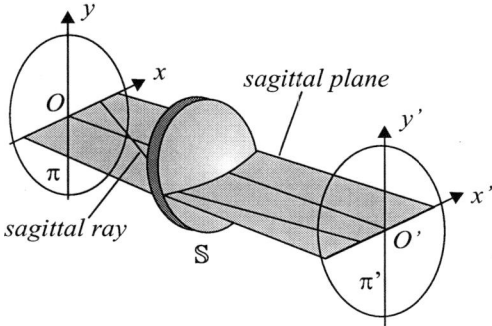

Fig. 1.7 Sagittal rays

Remark The above definition of an ideal system, which is based on geometric optics, is too restrictive since it does not take into account the wave character of light. In fact, as we shall see in Chapter 3, due to the wave nature of light propagation, the light rays from an object point $x \in \pi$ cannot converge at the same image point $x' \in \pi'$. They are actually spread in a complex way over a small circle, which is called the **Airy disk**. Consequently, the optical system will be perfect if the geometric spot produced by the rays from $x \in \pi$ is contained in the Airy disk. When this happens, the optical system is said to be **diffraction limited**. Finally, if the optical combination is designed for photographic use, the lowest limit of the geometric spot is represented by the grain of the film.

1.3 Axial Chromatic Aberrations

Let us suppose that the paraxial approximation can be applied to the optical system we are considering.

In this approximation, the dependence of the image plane π' on the wavelength λ results in **axial chromatic aberration**. If the image plane π'_b corresponding to blue light is closer to the optical system $\mathbb{S}$ than the image plane π'_r corresponding to red light, then the axial chromatic aberration is said to be **undercorrected**; in the opposite case it is **overcorrected**. Denoting the distances of π'_b and π'_r from the last surface of $\mathbb{S}$ by z'_b and z'_r, respectively, the difference

$$\text{PAC} = z'_b - z'_r, \tag{1.8}$$

is called the ***primary axial color***, abbreviated to PAC.

The optical system $\mathbb{S}$ is said to be **achromatic** if $\pi'_b \equiv \pi'_r$. Since the refractive index depends on the wavelength, generally speaking the image planes for different colors do not coincide. This circumstance causes the **secondary spectrum**. An optical system is said to be **apochromatic** when the image planes corresponding to three colors (for instance, blue, red, and green) coincide.

Another important paraxial chromatic aberration is the **primary lateral color**. This is related to the dependence of the magnification $M(\lambda)$ on the wavelength λ. Consequently, the size of the blue image is not equal to the size of the red image. This aberration is defined by the difference:

$$PLC = (M(\lambda_b) - M(\lambda_r))y, \qquad (1.9)$$

where λ_b and λ_r are the wavelengths of blue and red light, respectively, and y is the height of the object on the optical axis.

Other chromatic defects derive from the dependence of monochromatic aberrations on the wavelength. These aberrations will be discussed in the next section.

1.4 Monochromatic Aberrations

In this section, we suppose that:

- *The wavelength of the light is fixed*

- *The paraxial approximation is not verified*

- *The plane π' is the Gaussian image plane corresponding to the fixed wavelength.*

In order to define a ray from the object point (x, y), we can identify the cosines of the angles that it forms with the axes. Equivalently, if we introduce an auxiliary plane π_e,[1] the ray is defined by the coordinates (x, y) of the object point from which it originated as well as by the Cartesian coordinates (x_e, y_e) or the polar coordinates (r_e, φ_e) of the point at which it intersects π_e (see Figure 1.8). In general, the coordinates (x', y') of the point at which a ray from (x, y) meets the image plane π' depend on both the object point and the ray; that is:

$$x' = f_x(x, y, x_e, y_e) = g_x(x, y, r_e, \varphi_e),$$
$$y' = f_y(x, y, x_e, y_e) = g_y(x, y, r_e, \varphi_e).$$

---

[1] In what follows, π_e will correspond to the Gaussian entrance pupil of the optical system.

1.4. Monochromatic Aberrations

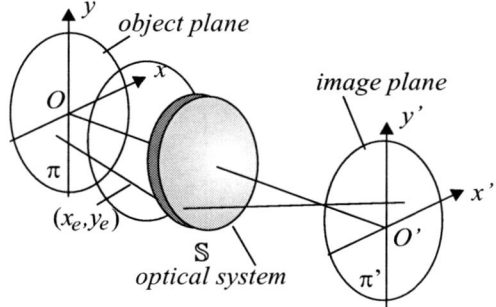

Fig. 1.8 Object plane, entrance pupil, and image plane

The functions

$$\epsilon_x = x' - Mx, \qquad (1.10)$$
$$\epsilon_y = y' - My, \qquad (1.11)$$

which describe the departure of the optical system from ideal behavior, are called **monochromatic aberration functions**.

It is very important to know these functions in order to control or eliminate the aberrations in an optical system. However, only the first terms of Taylor's expansion of functions (1.10) and (1.11) are known.[2] If the third-order powers of the independent variables are taken into account, Von Seidel–Schwarzschild's third-order aberration theory is obtained, whereas Buchdal's aberration theory also includes fifth-order powers. Third-order aberrations will be discussed in the next few chapters, together with many applications.

Another way to analyze a *given* optical system is through **ray tracing**. This method is easy to perform using today's powerful computers. Essentially, it involves computing the *exact* trajectories of rays through a *given* optical system $\mathbb{S}$ using the refraction law at any surface of $\mathbb{S}$. In this way, the point (x', y') at which any ray meets the image plane π' can be determined (see Section 1.6). Using this procedure it is possible to obtain very accurate representations of the aberrations. For instance, consider an object point $(0, y)$. If only meridional rays are analyzed, starting from $(0, y)$, then $\epsilon_x = 0$ and the aberration function ϵ_y depends only on the variables y and y_e (see Figure 1.9):

$$\epsilon_y = f_y(y, y_e) - My. \qquad (1.12)$$

---

[2]In the next few sections we prove that only odd powers of the variables x, y, x_e, y_e appear in Taylor's expansion of (1.10) and (1.11).

12 Chapter 1. Fermat's Principle and General Considerations

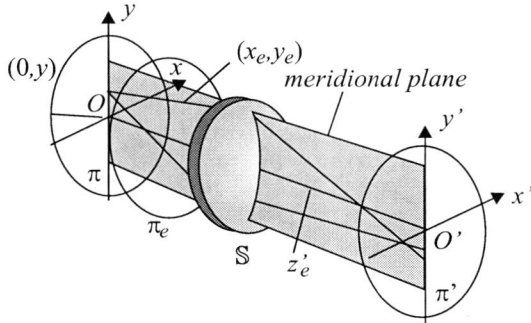

Fig. 1.9 Meridional rays, object plane, entrance pupil, and image plane

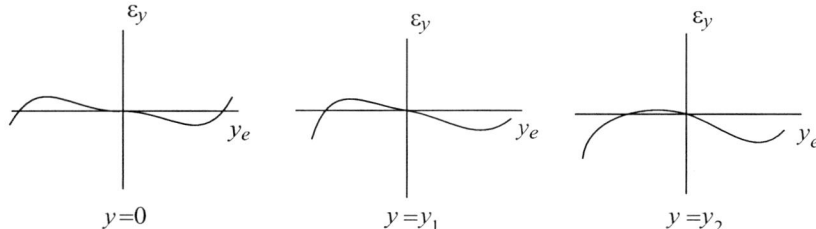

Fig. 1.10 Ray-intercept curves

This equation, for a fixed y, defines a curve in the plane y_e, ϵ_y called the **ray-intercept curve**, which supplies the displacements along $O'y'$ with respect to the ideal image $(0, My)$ of all meridional rays starting from the object point $(0, y)$. For an ideal system, it reduces to a straight line that passes through the origin with zero slope. This curve is usually traced for meridional rays starting from an object point on the optical axis and then from two other points $(0, y_1)$, $(0, y_2)$, where y_2 corresponds to the greatest ordinate of the object and y_1 to a fraction of y_2, usually $0.7y_2$ (see Figure 1.10).

If we consider the set of rays that come from $(0, y)$ and intercept the plane π_e along the Ox−axis, then the functions (1.10) reduce to:

$$\epsilon_x = f_x(y, x_e). \tag{1.13}$$

Again, for a fixed y, relation (1.13) defines a curve in the plane x_e, ϵ_x, which, due to the symmetry of the optical system, is symmetric with respect to the ϵ_x−axis. Thus, it is sufficient to represent it only in the region $x_e > 0$.

1.5. Surfaces of Revolution

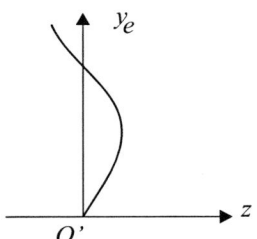

Fig. 1.11 Longitudinal spherical aberration curve

Also important is the **longitudinal spherical aberration curve**. Consider the rays that come from the object point $(0,0)$ on the optical axis and that are present in the meridional plane Oyz. Each of these rays, which cross π_e at $(0, y_e)$, meets the optical axis at a distance z' from the last surface of the system $\mathbb{S}$, which depends only on y_e. The longitudinal spherical aberration curve is simply the locus of all points (z', y_e) (see Figure 1.11).

It is important to note that all of these curves can be plotted for different colors.

Finally, if we choose a certain number of rays and some wavelengths, the set of points at which these rays intersect the image plane provides a useful representation of the aberrations of a given system; this is called a **spot diagram**.

1.5 Surfaces of Revolution

In this section the basic aspects of surfaces of revolution will be reviewed, since such surfaces are the type most frequently used in optical systems. A **surface of revolution** S is obtained by rotating a curve γ about an axis z, which is called the axis of symmetry of S. The intersection of S with any plane π containing z is a curve, and the equation of this curve always takes the same form, $z = z(\rho)$, where ρ is the distance of any point on γ from z (see Figure 1.12). Due to the symmetry of S, the curve γ is an even function of ρ:

$$z(\rho) = z(-\rho). \qquad (1.14)$$

We now suppose that S is obtained by rotating a **conic** γ about the axis z. There are many ways to define such a curve. First, a conic is obtained by intersecting a cone with a plane. It can also be defined as the locus of the

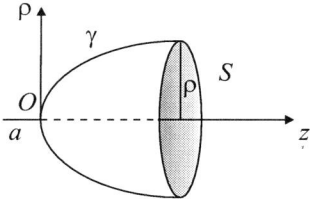

Fig. 1.12 Surface of revolution

points P such that the ratio e of their distances from a straight line a and a point $F \notin a$ is constant. The straight line a, the point $F \equiv (0, c)$, and the constant e are called the **directrix**, the **focus**, and the **eccentricity** of γ, respectively. If P is any point on γ, then we have (see Figure 1.13):

$$\frac{PF}{PH} = \frac{\sqrt{(z-c)^2 + \rho^2}}{|z-d|} = e > 0. \quad (1.15)$$

In particular, when $P \equiv O$, we have

$$e = \frac{c}{|d|} > 0 \Rightarrow d = -\frac{c}{e}. \quad (1.16)$$

Equation 1.15 can also be written in the form

$$z^2(1-e^2) - 2(c-de)z + c^2 - d^2e^2 + \rho^2 = 0,$$

and, taking into account (1.16), it can be reduced to the condition

$$(1-e^2)z^2 - 2c(1+e)z + \rho^2 = 0.$$

In turn, using the relation

$$c(1+e) = 2f, \quad (1.17)$$

this becomes

$$(1-e^2)z^2 - 4fz + \rho^2 = 0. \quad (1.18)$$

It is more convenient to introduce the **conic parameter**

$$K = -e^2 < 0,$$

which allows us to write (1.18) as follows:

$$(1+K)z^2 - 4fz + \rho^2 = 0. \quad (1.19)$$

1.5. Surfaces of Revolution

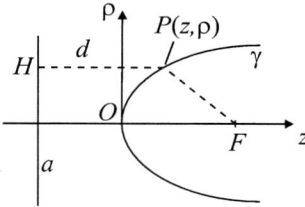

Fig. 1.13 A possible definition of a conic

In particular, when $K = -1$, we have the equation of a **parabola**

$$z = \frac{\rho^2}{4f}. \tag{1.20}$$

In the other cases, we have the equivalent explicit form

$$z = \frac{2f \mp \sqrt{4f^2 - \rho^2(1+K)}}{1+K}. \tag{1.21}$$

Through analysis it can be shown that the radius of curvature R of a curve $z = z(\rho)$ at the origin is expressed by the following relation:

$$\frac{1}{R} = \frac{|z^{(2)}(0)|}{(1+z'(0))^{3/2}}, \tag{1.22}$$

which, when applied to (1.21) yields

$$\frac{1}{R} = \frac{1}{2f} \Leftrightarrow f = \frac{R}{2}. \tag{1.23}$$

Finally, in terms of the radius of curvature R at the vertex, (1.20) and (1.21) assume the form

$$\begin{aligned} z &= \frac{\rho^2}{2R}, \\ z &= \frac{R \mp \sqrt{R^2 - \rho^2(1+K)}}{1+K}. \end{aligned} \tag{1.24}$$

Table 1.1 describes the classification scheme for quadric surfaces of revolution.

In the following we frequently use the following approximate form of a quadric surface of revolution, which is derived from Taylor's expansion of (1.24) when the minus sign is chosen:

$$z = \frac{\rho^2}{2R} + \frac{(1+K)\rho^4}{8R^3} + \frac{(1+K)^2\rho^6}{16R^5} + \cdots. \tag{1.25}$$

Parameter K	Eccentricity	Surface
$K = 0$	$e = 0$	Sphere
$-1 < K < 0$	$0 < e < 1$	Ellipsoid
$K = -1$	$e = 1$	Paraboloid
$K < -1$	$e > 1$	Hyperboloid

Table 1.1 Parameters and eccentricities of different conics

In the above discussion we considered a surface of revolution produced by rotating a conic about its axis of symmetry. Let $z = z(\rho)$ be any symmetric curve γ with respect to the Oz-axis for which the condition $z(0) = 0$ is valid. If we denote the surface of revolution obtained by rotating γ around Oz by S, and take into account the symmetry of γ with respect to Oz, then Taylor's expansion of S yields

$$z(\rho) = \frac{z^{(2)}(0)}{2!}\rho^2 + \frac{z^{(4)}(0)}{4!}\rho^4 + \frac{z^{(6)}(0)}{6!}\rho^6 + \cdots.$$

By introducing the radius of curvature R (see (1.22)) of the surface of revolution S at the origin, where $z'(0) = 0$:

$$R = \frac{1}{z^{(2)}(0)},$$

the previous expansion becomes

$$z(\rho) = \frac{\rho^2}{2R} + \frac{z^{(4)}}{4!}\rho^4 + \frac{z^{(6)}}{6!}\rho^6 + \cdots, \qquad (1.26)$$

with the usual notation $\rho = \sqrt{x^2 + y^2}$.

To facilitate a comparison between a quadric of revolution and any surface of revolution for which $R \neq 0$, the following relations are introduced

$$\begin{aligned}\frac{z^{(4)}}{4!} &= \frac{(1+K)}{8R^3}, \\ \frac{z^{(6)}}{6!} &= \frac{(1+K_1)^2}{16R^5},\end{aligned} \qquad (1.27)$$

which define the constants K and K_1. Using (1.27), expansion (1.26) assumes the form

$$z(\rho) = \frac{\rho^2}{2R} + \frac{(1+K)}{8R^3}\rho^4 + \frac{(1+K_1)^2}{16R^5}\rho^6 + \cdots. \qquad (1.28)$$

We conclude with the following remarks:

- Up to second order, all of the surfaces of revolution differ only by the radius of curvature at the vertex

- The paraboloid is exactly represented by the first term of expansion (1.28).

In some cases, it is more convenient to represent any surface of revolution by the equation

$$z = \frac{1}{2R}\rho^2 + a_4\rho^4 + \cdots. \tag{1.29}$$

1.6 Ray Tracing in Axially Symmetric Systems

All of the graphical representations of the aberrations described in Section 1.4 are based on following the path of a ray across the whole optical system $\mathbb{S}$.

The optical system $\mathbb{S}$ is assumed to contain n surfaces of revolution S_i, $i = 1, \ldots, n$. We denote by N_i and N_{i+1} the refractive indices of the media before and after S_i, respectively. Finally, t_i represents the distance between S_i and S_{i+1} along the optical axis (i.e., the distance between their vertices). A ***single*** Cartesian coordinate system $O_1 xyz$ that has its origin O_1 at the vertex of S_1 and the $O_1 z$-axis along the optical axis (see Figure 1.14) is introduced. We denote by z the distance of the object plane from S_1, provided that it is situated at a finite distance from S_1. Finally, w_e is the distance of the reference plane π_e (see Section 1.2)[3] from S_1 and z' denotes the distance of the image plane from the last surface S_n of $\mathbb{S}$.

Based on the results from the previous section, the equation $z = f_i(x, y)$ for any surface S_i is:

$$z = \sum_{j=1}^{i-1} t_j, \quad \text{if } S_i \text{ is planar},$$

$$z = \sum_{j=1}^{i-1} t_j + \frac{x^2 + y^2}{2R_i}, \quad \text{if } K_i = -1 \text{ (paraboloid)},$$

$$z = \sum_{j=1}^{i-1} t_j + \frac{R_i \mp \sqrt{R_i^2 - (1+K_i)(x^2+y^2)}}{1+K_i}, \quad \text{if } K_i \neq -1,$$

$$z = \sum_{j=1}^{i-1} t_j + a_{4,i}(x^2+y^2) + a_{6,i}(x^2+y^2)^2,$$

if S_i is any aspherical surface.

---

[3] As we said before in Section 1.2, this plane will correspond to the entrance plane of the system (see Section 2.4).

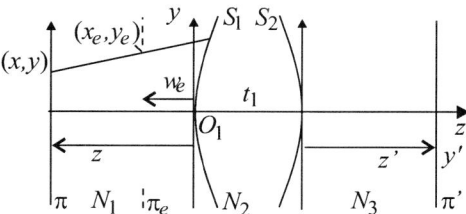

Fig. 1.14 Notation adopted

In the third equation, we choose the minus sign when $R_i > 0$ and the plus sign in the opposite case.

In order to follow the path of a ray Γ across $\mathbb{S}$, we must first determine its equations before it intersects with the first surface S_i of $\mathbb{S}$. If the object is situated at a finite distance from the first surface of the optical system $\mathbb{S}$, the object point (x, y) can be identified together with the point on the reference plane π_e intercepted by the ray. Based on these quantities, we evaluate the director cosines α_1, β_1, and γ_1 of Γ, oriented according to the direction of propagation of the light:

$$\alpha_1 = \frac{x_e - x}{\sqrt{(x_e - x)^2 + (y_e - y)^2 + (w_e - z)^2}},$$

$$\beta_1 = \frac{y_e - y}{\sqrt{(x_e - x)^2 + (y_e - y)^2 + (w_e - z)^2}}, \quad (1.30)$$

$$\gamma_1 = \frac{w_e - z}{\sqrt{(x_e - x)^2 + (y_e - y)^2 + (w_e - z)^2}}.$$

If the object is at infinity, Γ is determined by the director cosines α_1, β_1, and γ_1 and the coordinates (x_e, y_e).

In both cases, the parametric equations of the ray Γ are

$$\xi_1 = x_e + \alpha_1 s,$$
$$\eta_1 = y_e + \beta_1 s, \quad (1.31)$$
$$\zeta_1 = w_e + \gamma_1 s,$$

where (ξ_1, η_1, ζ_1) are the coordinates of the moving point on Γ and s is a parameter that coincides with the distance between (x_e, y_e, w_e) and (ξ_1, η_1, ζ_1). Recall that the director cosines are not independent since they satisfy the condition $\alpha_1^2 + \beta_1^2 + \gamma_1^2 = 1$.

Now, in order to determine the point of intersection $P_1 \equiv (x_1, y_1, z_1)$ between Γ and S_1, the following equation must be solved:

$$w_e + \gamma_1 s = f_1(x_e + \alpha_1 s, y_e + \beta_1 s), \quad (1.32)$$

1.6. Ray Tracing in Axially Symmetric Systems

which supplies the value of s corresponding to P_1. This equation is not linear, so we apply Newton's method in order to solve it. However, this method requires that we know an approximate value of the solution. To this end, it is sufficient to consider the point of intersection P_1^* of the ray with the tangent plane π_1^* to S_1 at its vertex. The value s_1^* of the parameter corresponding to P_1^* is obtained by inserting $\zeta_1 = 0$ into the third equation of (1.31) and solving the resulting equation with respect to s:

$$s_1^* = -\frac{w_e}{\gamma_1}. \tag{1.33}$$

Starting from this value of the parameter, it is possible to determine the solution s_1 of (1.32) by Newton's method. Then, the coordinates of P_1 yield

$$\begin{aligned} x_1 &= x_e + \alpha_1 s_1, \\ y_1 &= y_e + \beta_1 s_1, \\ z_1 &= w_e + \gamma_1 s_1, \end{aligned} \tag{1.34}$$

In order to follow the ray Γ across $\mathbb{S}$, it is sufficient to prove that—assuming the equations of Γ before the surface S_i of $\mathbb{S}$ are known together with the point $P_i \equiv (x_i, y_i, z_i) \in S_i$ at which it intersects S_i—it is possible to find the parametric equations of the ray Γ arriving at $P_{i+1} \equiv (x_{i+1}, y_{i+1}, z_{i+1}) \in S_{i+1}$.

First, the refraction law at P_i must be applied. To this end, we consider the components of the unit normal to S_i at P_i, oriented in the direction of propagation of the light:

$$\begin{aligned} n_{i,x} &= -\mu_i \frac{\partial f_i}{\partial x}, \\ n_{i,y} &= -\mu_i \frac{\partial f_i}{\partial y}, \\ n_{i,z} &= \mu_i, \end{aligned} \tag{1.35}$$

where

$$\mu_i = \frac{1}{\sqrt{1 + \left(\frac{\partial f_i}{\partial x}\right)^2 + \left(\frac{\partial f_i}{\partial y}\right)^2}}. \tag{1.36}$$

The following quantities can now be determined at P_i (see (1.2) and (1.3)):

$$\cos i_i = n_{i,x}\alpha_i + n_{i,y}\beta_i + n_{i,z}\gamma_1, \quad \sin i_i = \sqrt{1 - \cos^2 i_i},$$

$$\lambda_i = N_{i+1}\sqrt{1 - \frac{N_i^2}{N_{i+1}^2}\sin^2 i_i} - N_i \cos i_i, \tag{1.37}$$

and the director cosines of the refracted ray become:

$$\alpha_{i+1} = \frac{1}{N_{i+1}}(N_i\alpha_i + \lambda_i n_{i,x}),$$
$$\beta_{i+1} = \frac{1}{N_{i+1}}(N_i\beta_1 + \lambda_i n_{i,y}), \qquad (1.38)$$
$$\gamma_{i+1} = \frac{1}{N_{i+1}}(N_i\gamma_i + \lambda_i n_{i,z}).$$

Consequently, the parametric equations of the refracted ray are

$$\xi_{i+1} = x_i + \alpha_{i+1}s,$$
$$\eta_{i+1} = y_i + \beta_{i+1}s, \qquad (1.39)$$
$$\zeta_{i+1} = z_i + \alpha_{i+1}s.$$

This ray meets the tangent plane π_{i+1}^* to S_{i+1} at its vertex when the parameter a has the value

$$s^* = \frac{1}{\alpha_{i+1}}(\sum_{j=1}^{i} t_j - z_i). \qquad (1.40)$$

The procedure that we have described can be iterated starting from these last equations.

1.7 Exercises

1. Show that Fermat's principle implies the refraction law.

   Let $\mathbf{x}_0$ and $\mathbf{x}_1$ be two arbitrary points in two media with constant refractive indices N and N', respectively. Moreover, let S denote a regular surface that separates these two media. According to Fermat's principle, any ray from $\mathbf{x}_0$ to $\mathbf{x}_1$ follows a rectilinear path γ_0 from $\mathbf{x}_0$ to a point $\mathbf{x} \in S$, and another rectilinear path γ_1 from $\mathbf{x}$ to $\mathbf{x}_1$. Moreover, the point $\mathbf{x}$ must be determined under the condition that the optical path length OPL along $\gamma_0 \bigcup \gamma_1$ is stationary. Consequently, if $F(\mathbf{x}) = 0$ denotes the equation of S and

   $$G(\mathbf{x}) = N\sqrt{\sum_{i=1}^{3}(x_i - x_{0i})^2} + N'\sqrt{\sum_{i=1}^{3}(x_{1i} - x_i)^2} \qquad (1.41)$$

is the optical path length, we must minimize the function
$$G + \lambda F,$$
where λ is a suitable Lagrangian multiplier. Thus, we have the equation
$$\nabla G + \lambda \nabla F = \mathbf{0},$$
which can also be written as follows:
$$\nabla G - \lambda \mathbf{n}|\nabla F| \equiv \nabla G + \mu \mathbf{n} = \mathbf{0}, \tag{1.42}$$
where $\mathbf{n}$ is the internal normal to S and μ is another Lagrangian multiplier.

Since this relation implies
$$\mu = -\nabla G \cdot \mathbf{n}, \tag{1.43}$$
the condition (1.42) can also be written in the form
$$\nabla G - (\nabla G \cdot \mathbf{n})\mathbf{n} = \mathbf{0}, \tag{1.44}$$
and the following relation is obtained:
$$\nabla G \times \mathbf{n} = \mathbf{0}. \tag{1.45}$$
On the other hand, from (1.41) we can easily derive the relation
$$\nabla G = N\mathbf{t} - N'\mathbf{t}', \tag{1.46}$$
where $\mathbf{t}$ and $\mathbf{t}'$ are the unit vectors along γ_0 and γ_1, respectively. Results (1.46) and (1.45) can be placed in the form
$$\begin{aligned} N\mathbf{t} - N'\mathbf{t}' &= -\mu \mathbf{n}, \\ N\mathbf{t} \times \mathbf{n} &= N'\mathbf{t}' \times \mathbf{n}. \end{aligned} \tag{1.47}$$
The first of these relations shows that γ_0, γ_1 and $\mathbf{n}$ are present in the same plane, whereas the second one implies the refraction law.

2. Verify that the foci of an ellipsoidal mirror are stigmatic points.

Consider a reflecting surface S that is an ellipsoid of revolution about an axis a. Let A and A' be its foci. If P is the point at which the ray from $\mathbf{x}$ meets the mirror, then, according to the definition of ellipsoid of revolution around a, we have (see Figure 1.15)
$$AP + PA' = \text{const.} \tag{1.48}$$
The optical path length of any such a path between A and A' is thus independent of P. Consequently, these paths are rays (see Section 1.1), and A and A' are stigmatic points.

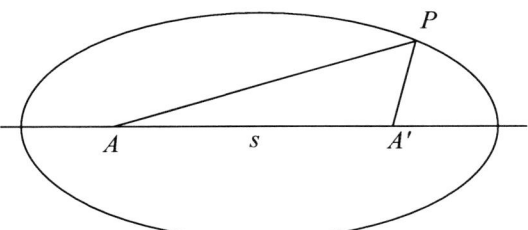

Fig. 1.15 Reflecting ellipsoid

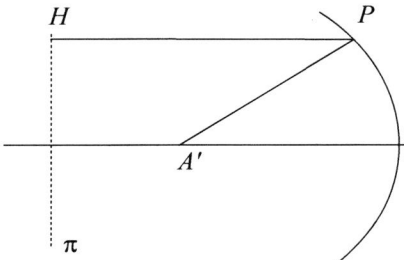

Fig. 1.16 Reflecting paraboloid

3. Let S be a mirror whose surface is a paraboloid of revolution. Verify that all of the rays parallel to the axis of revolution a of S intercept a at the focus A of S (see Figure 1.16).

   It is sufficient to consider a plane π orthogonal to a and to denote by H the point at which a ray meets π. If the definition of a paraboloid of revolution around a is taken into account:

   $$HP + PA' = \text{const.}, \qquad (1.49)$$

   we can conclude that, along any ray parallel to a, the optical path length from H to the focus a is constant.

4. Verify that the foci of an hyperbolic mirror are stigmatic points.

   The reflecting surface is a hyperboloid of revolution, and A and A' are its foci (see Figure 1.17). Therefore,

   $$AP - PA' = \text{const.} \qquad (1.50)$$

   and the foci are stigmatic points.

1.7. Exercises

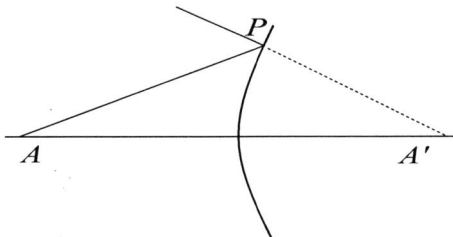

Fig. 1.17 Reflecting hyperboloid

5. Let S be a refractive surface of revolution separating two media with refractive indices N and N', and let a be the axis of S. Verify that two points A and A' on a, which are both located finite distances from S, are stigmatic if

$$N\,AP + N'\,PA' = \text{const.} \tag{1.51}$$

The corresponding refractive surface of revolution S is said to be an ovoid.

Chapter 2

Gaussian Optics

2.1 Gaussian Approximation for a Single Surface

In this chapter we present the Gaussian analysis of an optical system $\mathbb{S}$. In particular, we introduce all of the optical characteristics of $\mathbb{S}$ (pupils, focal points, nodal points, principal planes, etc.). All these data can be obtained using the notebook *GlobalGaussianData*.

Among the many potential ways of presenting this topic, we use Fermat's principle in order to introduce an approach that will be extended to third-order aberration theory.[1]

Let $\mathbb{S}$ be a single surface S with a symmetry of revolution about an axis a. N and N' denote the (constant) refractive indices of the media before and after S. Finally, let $Oxyz$ be a Cartesian frame of reference with $Oz \equiv a$ and the origin O located at the vertex of S (see Figure 2.1).

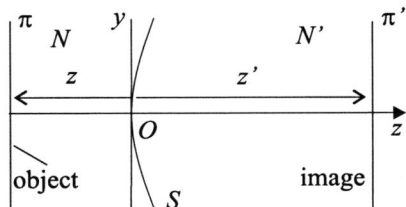

Fig. 2.1 Notation for a single surface

Note that Oyz is the meridional plane, whereas Oxz is the sagittal plane (see Section 1.2).

---
[1] Another elementary way of introducing paraxial optics is shown in Exercise 6 at the end of this chapter.

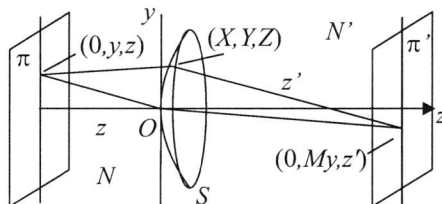

Fig. 2.2 Principal and marginal rays

The object is assumed to lie in a plane π. Let π' be a plane in the image space and P be a point on π. Due to the symmetry of S, it is always possible to choose P to be on the axis Oy, so it will have coordinates $(0, y, z)$. Consider the bundle Γ of polygonal lines γ that, starting from $(0, y, z)$, intersect S at the the point (X, Y, Z) and the plane π' at $(0, My, z')$ (see Figure 2.2). In particular, the ray γ_0, which originates at $(0, y, z)$ and intersects S at its vertex and the plane π' at $(0, My, z')$, belongs to Γ. We wish to know the conditions under which Γ is formed from *rays* or, equivalently, the points $(0, y, z)$ and $(0, My, z')$ are stigmatic (see Section 1.1).

Due to Fermat's principle, this happens if and only if the optical paths along the polygonal lines of Γ have the same value, or, equivalently, if the following optical path difference vanishes:

$$\Phi \equiv N\sqrt{X^2 + (Y-y)^2 + (Z-z)^2}$$
$$+ N'\sqrt{X^2 + (My-Y)^2 + (z'-Z)^2} \qquad (2.1)$$
$$- N\sqrt{y^2 + z^2} - N'\sqrt{M^2 y^2 + z'^2} = 0.$$

It is very simple to verify that the quantities

$$\xi = y^2, \quad \eta = X^2 + Y^2, \quad \zeta = yY, \qquad (2.2)$$

are rotational invariants (i.e., that they are invariant under an arbitrary rotation of the plane Oxy about the Oz-axis). Introducing (2.2) into (2.1), we obtain:

$$\Phi \equiv N\sqrt{z^2 + \eta - 2\zeta + \xi - 2zZ + Z^2}$$
$$+ N'\sqrt{z^2 + \eta - 2M\zeta + M^2\xi - 2zZ + Z^2} \qquad (2.3)$$
$$- N\sqrt{z^2 + \xi} - N'\sqrt{z'^2 + M^2\xi} = 0.$$

This condition is analyzed here using the **Gaussian or paraxial approximation**, which is defined as follows.

The optical path difference Φ among all the optical paths $\gamma \in \Gamma$ and γ_0 is constant to within second-order terms in the variables

2.1. Gaussian Approximation for a Single Surface

ξ, η and ζ, or, equivalently, to within fourth-order terms in the variables $y, X,$ and Y.

In other words, we are supposing that the angles that all of the rays that intersect with the surface S form with the optical axis a, as well as the distances of the object points from a, comply with the following condition: that the coordinates (y, X, Y) of all the points at which the rays intersect wih the base-planes π and π' and the surface S are so small that, to a first approximation, it is possible to neglect the fourth-order powers of y, X, and Y in the optical paths.

In order to determine the consequences of (2.3), we start by noting that, due to the symmetry of revolution of S, the equation $Z = F(X,Y)$ of S necessarily takes the form

$$Z = F(X^2 + Y^2).$$

Consequently, in the Gaussian approximation, it can be written as follows (see Section 1.5):

$$Z = \frac{1}{2R}(X^2 + Y^2), \qquad (2.4)$$

where R denotes the radius of curvature of S at its vertex O.

In view of Taylor's expansion

$$\sqrt{a^2 + x} = a + \frac{x}{2a} + O(2), \qquad (2.5)$$

from (2.3) we can derive

$$\Phi = \frac{1}{2}\left(\frac{N}{R} - \frac{N}{z} - \frac{N'}{R} + \frac{N'}{z'}\right)\eta + \left(\frac{N}{z} - M\frac{N'}{z'}\right)\zeta + O(2) = 0. \qquad (2.6)$$

It is evident that the first-order terms vanish for any value of ξ, η, and ζ if and only if

$$Q \equiv \frac{N}{R} - \frac{N}{z} = \frac{N'}{R} - \frac{N'}{z'}, \qquad (2.7)$$

$$M \equiv \frac{y'}{y} = \frac{Nz'}{N'z}. \qquad (2.8)$$

The following conclusions can be derived from the above equations:

1. The image of an object point along the axis Ox is located along the axis $O'y'$ for any (X, Y). Consequently, in this approximation, we only need to consider rays in this plane in order to obtain the image of a given object.

2. Relation (2.7) supplies the position of the image plane π' with respect to the surface S when S and the position of the object plane π are given. The quantity Q is called **Abbe's invariant**. The planes π and π' are said to be *conjugate* when z and z' satisfy (2.7).

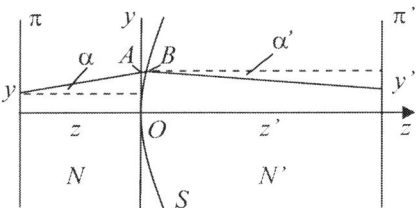

Fig. 2.3 In the Gaussian approximation $A \equiv B$

3. The **magnification factor** (2.8) is independent of y, meaning that the image is proportionally reduced if $|M| < 1$ and enlarged if $|M| > 1$. When $M < 0$, the image is inverted with respect to the object.

We conclude this section by showing that, using the above results, it is possible to deduce the correspondence that exists between the angles α and α', which are the angles that a ray originating from an object point on y form with the optical axis before and after refraction, respectively. The angles are taken to be positive if the oriented optical axis coincides with the oriented ray after a counterclockwise rotation, and negative in the opposite case.

In the Gaussian approximation, the points A and B in Figure 2.3 coincide, and we have the relation[2]

$$y - z\alpha = y' - z'\alpha', \qquad (2.9)$$

which, in view of (2.8), can also be written as follows:

$$\alpha' = \frac{1}{N'}\left(\frac{N}{z} - \frac{N'}{z'}\right)y + \frac{z}{z'}\alpha. \qquad (2.10)$$

Taking into account (2.7), we finally obtain

$$\alpha' = -\frac{P}{N'}y + \frac{z}{z'}\alpha, \qquad (2.11)$$

where

$$P = \frac{N' - N}{R}, \qquad (2.12)$$

is the **power** of the surface S.

---
[2] Note that in Figure 2.3, $z < 0$, $\alpha > 0$, $z' > 0$, $\alpha' < 0$.

Equations 2.8 and 2.11 lead to the following system:

$$y' = \frac{Nz'}{N'z}y,$$
$$\alpha' = -\frac{P}{N'}y + \frac{z}{z'}\alpha. \qquad (2.13)$$

Remark If θ and θ' denote the angles that the ray starting from (y, z) and reaching the vertex O of S forms with the optical axis before and after the refraction, the following relations hold

$$y = \theta z, \quad y' = \theta' z',$$

and the equations (2.13) can equivalently be written as follows:

$$y' = \frac{Nz'}{N'}\theta, \quad \theta' = \frac{N}{N'}\theta. \qquad (2.14)$$

When the object is at infinity, (2.13) become meaningless and they can be replaced by (2.14). The corresponding value of z' is called the **posterior focal length** of the surface. From (2.7), we can derive

$$f' = \frac{N'}{P}, \qquad (2.15)$$

and the first equation of (2.14) yields

$$y' = \frac{N}{P}\theta. \qquad (2.16)$$

In particular, if S is a mirror, $N = -N' = 1$, $P = -2/R$, and we have

$$f' = \frac{R}{2}, \quad y' = -\frac{R}{2}\theta. \qquad (2.17)$$

2.2 Compound Systems

Let $\mathbb{S}$ be an optical system containing n surfaces of revolution $S_1 \ldots S_n$ with the same optical axis a. The distance and the refractive index between the surface S_i and the next surface S_{i+1} will be denoted by t_i and N'_i, respectively. Finally, z_1 and z'_n represent the distances of the object plane π from S_1 and the distance of the image plane π' from S_n (see Figure 2.4).

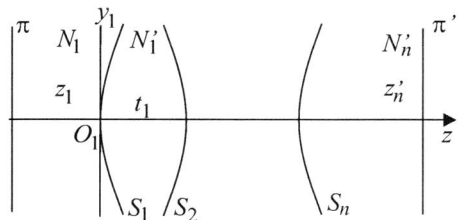

Fig. 2.4 Notation for a compound system

By applying (2.7) and (2.13) to each surface S_i, we get

$$\frac{N_i}{R_i} - \frac{N_i}{z_i} = \frac{N'_i}{R_i} - \frac{N'_i}{z'_i}, \tag{2.18}$$

$$y'_i = \frac{N_i z'_i}{N'_i z_i} y_i, \tag{2.19}$$

$$\alpha'_i = -\frac{P_i}{N'_i} y_i + \frac{z_i}{z'_i} \alpha_i, \tag{2.20}$$

and using the evident equations

$$y_{i+1} = y'_i, \tag{2.21}$$
$$z_{i+1} = z'_i - t_i, \tag{2.22}$$
$$\alpha_{i+1} = \alpha'_i, \tag{2.23}$$
$$N_{i+1} = N'_i, \tag{2.24}$$

it is possible to evaluate the path of any ray through the whole system $\mathbb{S}$.

Although the above formulae allow us to determine the image plane and the size of the image of any object in the Gaussian approximation, it is interesting to rewrite them in a form that gives prominence to some fundamental properties of a compound system $\mathbb{S}$. First, we write relations (2.19), (2.20) in the matrix form

$$\begin{pmatrix} y'_i \\ \alpha'_i \end{pmatrix} = \mathbb{T}_i \begin{pmatrix} y_i \\ \alpha_i \end{pmatrix}, \tag{2.25}$$

where

$$\mathbb{T}_i = \begin{pmatrix} \dfrac{N_i z'_i}{N'_i z_i} & 0 \\ -\dfrac{P_i}{N'_i} & \dfrac{z_i}{z'_i} \end{pmatrix}. \tag{2.26}$$

2.2. Compound Systems

Moreover, the following identity can be verified (see Exercise 1):

$$\mathbb{T}_i = \mathbb{D}_i^l \mathbb{R}_i \mathbb{D}_i^r, \tag{2.27}$$

where

$$\mathbb{D}_i^l = \begin{pmatrix} 1 & z_i' \\ 0 & 1 \end{pmatrix}, \quad \mathbb{D}_i^r = \begin{pmatrix} 1 & -z_i \\ 0 & 1 \end{pmatrix}, \tag{2.28}$$

and

$$\mathbb{R}_i = \begin{pmatrix} 1 & 0 \\ -\dfrac{P_i}{N_i'} & \dfrac{N_i}{N_i'} \end{pmatrix}. \tag{2.29}$$

It is evident that the matrix $\mathbb{D}_i^r$ represents the linear transformation

$$y_i^* = y_i - z_i \alpha_i, \quad \alpha_i^* = \alpha_i, \tag{2.30}$$

determined by the rectilinear rays between the object plane π_i of S_i and the tangent plane π_i^* to S_i at its vertex. A similar meaning must be attributed to the matrix $\mathbb{D}_i^l$, which refers to the corresponding transformation between π_i^* and the image plane π_i' of S_i. In other words, in the decomposition (2.27), the linear transformation between π_i and π_i' is regarded as the combination of the following linear transformations:

1.

$$y_i^* = y_i - z_i \alpha_i,$$
$$\alpha_i^* = \alpha_i,$$

   between π_i and π_i^*,

2.

$$y_i^{*\prime} = y_i^*,$$
$$\alpha_i^* = -\dfrac{P_i}{N_i'} y_i^* + \dfrac{N_i}{N_i'} \alpha_i^*,$$

   between π_i^* and π_i^* itself, which takes into account the refraction that occurs at the surface S_i, and

3.

$$y_i' = y_i^{*\prime} - z_i' \alpha_i^{*\prime},$$
$$\alpha_i' = \alpha_i^{*\prime},$$

   between π_i^* and π_i'.

The above considerations allow us to conclude that the total transformation between π_1 and π'_n generated by the whole system $\mathbb{S}$ is described by the linear transformation

$$\begin{pmatrix} y'_n \\ \alpha'_n \end{pmatrix} = \begin{pmatrix} 1 & z'_n \\ 0 & 1 \end{pmatrix} \mathbb{M} \begin{pmatrix} 1 & -z_1 \\ 0 & 1 \end{pmatrix} \begin{pmatrix} y_1 \\ \alpha_1 \end{pmatrix}, \tag{2.31}$$

where

$$\mathbb{M} = \mathbb{R}_n \mathbb{D}^r_n \mathbb{D}^l_{n-1} \cdots \mathbb{D}^r_2 \mathbb{D}^l_1 \mathbb{R}_1. \tag{2.32}$$

The expression of $\mathbb{M}$ can be simplified by noting that, using (2.22) and (2.28), we have

$$\mathbb{D}_i \equiv \mathbb{D}^r_{i+1} \mathbb{D}^l_i = \begin{pmatrix} 1 & t_i \\ 0 & 1 \end{pmatrix}. \tag{2.33}$$

This matrix describes the linear transformation between the tangent planes π^*_i and π^*_{i+1} and the surfaces S_i and S_{i+1} at their vertices. In view of this result, we write the matrix $\mathbb{M}$ in form

$$\mathbb{M} = \mathbb{R}_n \mathbb{D}_{n-1} \cdots \mathbb{D}_1 \mathbb{R}_1 = \begin{pmatrix} M_{11} & M_{12} \\ M_{21} & M_{22} \end{pmatrix}, \tag{2.34}$$

so that $\mathbb{M}$ is the combination of all of the transformations (refractions and translations) inside the system $\mathbb{S}$, and so *it depends only on the characteristics of* $\mathbb{S}$. From (2.34), (2.33), and (2.29), it follows that

$$\det M = 1. \tag{2.35}$$

2.3 Principal Planes and Focal Lengths

In this section and the next, the matrix formulation of the linear transformation between the object and image planes is shown to be very useful for defining some global properties of the optical system $\mathbb{S}$.

First, the matrix equation (2.31) explicitly gives

$$\begin{aligned} y'_n &= (M_{11} + M_{21} z'_n) y_1 \\ &\quad + (-M_{11} z_1 + M_{12} - M_{21} z_1 z'_n + M_{22} z'_n) \alpha_1, \end{aligned} \tag{2.36}$$

$$\alpha'_n = M_{21} y_1 + (M_{22} - M_{21} z_1) \alpha_1. \tag{2.37}$$

The plane π_1 (whose distance from S_1 is z_1) and the plane π'_n (whose distance from S_n is z'_n) are conjugate if y'_n is independent of α_1; in other words, if the following condition of stigmatism is valid:

$$M_{12} - M_{11} z_1 - M_{21} z_1 z'_n + M_{22} z'_n = 0, \tag{2.38}$$

2.3. Principal Planes and Focal Lengths

When the optical system $\mathbb{S}$ (i.e., the matrix $\mathbb{M}$) and the distance z_1 are given, the value of z'_n corresponding to the conjugate plane π'_n can be obtained:

$$z'_n = -\frac{M_{11}z_1 - M_{12}}{M_{21}z_1 - M_{22}}. \tag{2.39}$$

Moreover, the quantity

$$M = M_{11} + M_{21}z'_n = \frac{1}{M_{22} - M_{21}z_1} \tag{2.40}$$

gives the total magnification of $\mathbb{S}$ with respect to the conjugate planes π and π'_n. For the last equality, we used (2.35) and (2.39).

In particular, from (2.39) we have:

$$\lim_{z_1 \to \infty} z'_n = -\frac{M_{11}}{M_{21}} \equiv f'_b, \tag{2.41}$$

and the quantity f_b is called the **back focal** of $\mathbb{S}$. The **front focal** f_f can be defined in a similar way.

Two conjugate planes π_p and π'_p are called the **anterior principal plane** and the **posterior principal plane** if the magnification related to them is equal to 1:

$$M_{11} + M_{21}z'_n = 1. \tag{2.42}$$

From conditions (2.39), (2.42), and (2.35), we obtain the following expressions for the distances z_p and z'_p of π_p and π'_p from the surfaces S_1 and S_n, respectively:

$$z_p = \frac{M_{22} - 1}{M_{21}}, \quad z'_p = \frac{1 - M_{11}}{M_{21}}. \tag{2.43}$$

Moreover, the quantities (see (2.43) and (2.41))

$$f = f_f - z_p, \quad f' = f'_b - z'_p = -\frac{1}{M_{21}} \tag{2.44}$$

are the **anterior focal length** and the **posterior focal length**.

Finally, the **nodal points** are two *conjugate* points on the optical axis a ($y_1 = 0$) such that any ray that intercepts one of them at an angle α to a also intercepts the other one at the same angle to a. From (2.36) and (2.37), the values z_{nd} and z'_{nd}, which define these two points, are given by the system

$$z'_{nd} = -\frac{M_{11}z_{nd} - M_{12}}{M_{21}z_{nd} - M_{22}}, \quad M_{22} - M_{21}z_{nd} = 1. \tag{2.45}$$

In the next section we show that the nodal points coincide with the intersections of the principal planes with a if $N_1 = N'_n$.

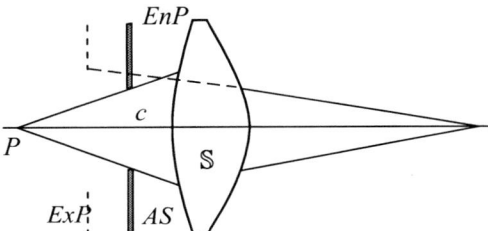

Fig. 2.5 Stop before the system

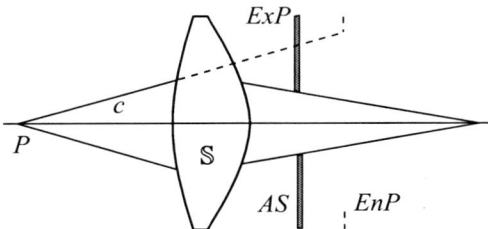

Fig. 2.6 Stop after the system

2.4   Stops and Pupils

Predicting the Gaussian position and the size of the image produced by an optical system $\mathbb{S}$ is not the only important task of paraxial analysis. It is equally important to establish the brightness of the image and the size of the field of view. These characteristics of the image are strongly conditioned by the presence of obstacles in the optical system as well as by the existence of lens rims, which control both the angular and spatial extensions of the light beams.

Consider an object point P located on the optical axis of the system $\mathbb{S}$, and the collection c of all the rays emitted from P that pass through $\mathbb{S}$. The **aperture stop** AS is the stop that effectively controls the angular extension of c and consequently determines the amount of light arriving at the image. Figures 2.5 and 2.6 show two positions for the aperture stop.

Let K^a be the part of the optical system $\mathbb{S}$ that is situated before the aperture stop; in other words, the part through which the light passes before meeting the aperture stop.

The image EnP of AS formed by K^a is called the **entrance pupil** of $\mathbb{S}$. Similarly, let K^b denote the part of $\mathbb{S}$ that the light passes through

2.4. Stops and Pupils

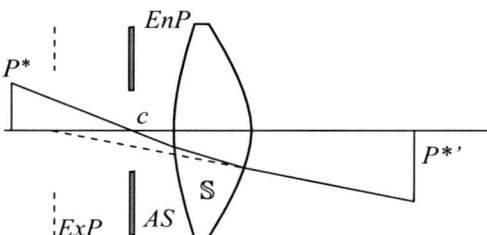

Fig. 2.7 Principal ray when the stop is positioned before the system

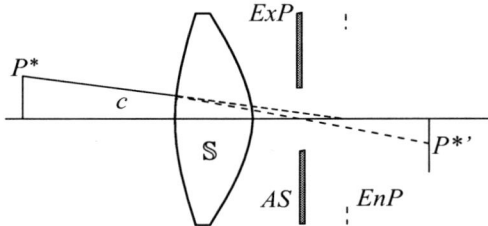

Fig. 2.8 Principal ray when the stop is positioned after the system

after AS. The image ExP of AS formed by K^b is called the **exit pupil** (see Figures 2.5 and 2.6 and the exercises at the end of the chapter). A ray from an off-axis object P^* through the center of the entrance pupil is called a **principal ray**. Since AS, EnP and ExP are conjugate, this principal ray will intercept the optical axis at the centers of the pupils and at the aperture stop.

Figures 2.7 and 2.8 show the path of the principal ray for the cases depicted in Figures 2.5 and 2.6. The presence of the aperture stop usually results in a different distribution of light on the image surface. This phenomenon is called **vignetting**. Because of it, the intensity of the light decreases from the axis to the periphery of the image. Figure 2.9 shows why vignetting occurs.

As the off-axis object point P^* is shifted away from the optical axis, the angular extension of the beam c' limited by the aperture stop reduces; moreover, this reduced amount of light is distributed over the same surface S, which is not orthogonal to the axis of c'.

Upon increasing the distance of P^* from the axis still further, some of the beam c' is intercepted by the rim of the lens in the optical system, and so the light arriving at the image is further reduced.

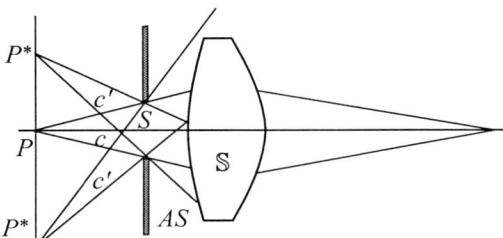

Fig. 2.9 Vignetting

2.5 Some Gaussian Optical Invariants

It is easy to verify that (see Exercise 2) the product $\mathbb{T}$ of the matrices $\mathbb{T}_i$ defined by (2.26) necessarily takes the form

$$\mathbb{T} = \begin{pmatrix} M & 0 \\ A & \dfrac{N_1}{N'_n}\dfrac{1}{M} \end{pmatrix}, \qquad (2.46)$$

where A is a suitable matrix. The transformation between the conjugate planes π_1 and π'_n then becomes

$$y'_n = My_1, \quad \alpha'_n = Ay_1 + \frac{N_1}{N'_n}\frac{1}{M}\alpha_1. \qquad (2.47)$$

When these relations refer to the conjugate principal planes, $M = 1$. Moreover, the ray intersecting the anterior principal plane at $y_1 = 0$ has a corresponding ray that crosses the posterior principal plane on the optical axis ($y'_n = 0$) and forms an angle with it of

$$\alpha'_n = \frac{N_1}{N'_n}\alpha_1.$$

Consequently, *if $N_1 = N'_n$, the intersection points of the principal planes with a are also nodal points.*

Some important Gaussian invariants will now be derived. Let π, π' and π_0, π'_0 be two pairs of conjugate planes. Due to (2.47), the point $y_1 \in \pi$ and the ray $(0, \alpha_1)$ originating from π (see Figure 2.10) will be transformed by $\mathbb{S}$ into the point

$$y'_n = My_1,$$

2.5. Some Gaussian Optical Invariants

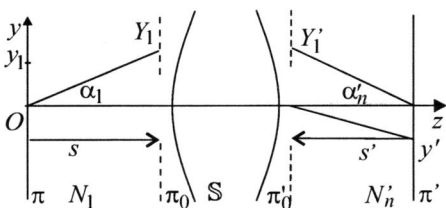

Fig. 2.10 Optical invariants

on π' and into the ray that intersects with π' on the optical axis and forms the following angle with it:
$$\alpha'_n = \frac{N_1}{N'_n M}\alpha_1.$$

From these equations we obtain **Helmholtz's invariant**:
$$N'_n y'_n \alpha'_n = N_1 y_1 \alpha_1, \tag{2.48}$$

which, in our approximation and resorting to the notations of Figure 2.10, can also be written
$$\frac{N'_n y'_n Y'_1}{s'} = \frac{N_1 y_1 Y_1}{s}. \tag{2.49}$$

We can derive three important consequences from (2.49):

- First, if the conjugate planes π_0 and π'_0 coincide, respectively, with the entrance and the exit pupils π_e and π'_e *of the whole system $\mathbb{S}$ or of a part of it*, we have $Y_1 = y_e$, $Y'_1 = M_e y_e$. Using (2.49), we can then derive
$$\frac{y'_n}{s'} = \frac{N_1}{N'_n}\frac{1}{M_e}\frac{y_1}{s}, \tag{2.50}$$
where M_e is the magnification related to the pair π_e and π'_e. Denoting the angle that the principal ray forms with the optical axis by θ and the corresponding angle in the exit pupil by θ', we have $y_1 = -s\theta$, $y'_n = -s'\theta'$, and so the previous relation assumes the form
$$\theta' = \frac{N_1}{N'_n}\frac{1}{M_e}\theta. \tag{2.51}$$

- Moreover, if π_0 and π'_0 coincide, respectively, with the anterior and posterior principal planes π_p and π'_p *of the whole system $\mathbb{S}$*, the magnification related to this pair of planes is equal to unity, and (2.50) implies that
$$\frac{s}{y_1} = \frac{N_1}{N'_n}\frac{s'}{y'_n}. \tag{2.52}$$

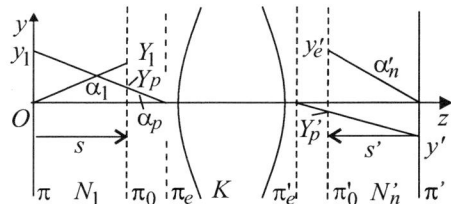

Fig. 2.11 Optical invariants for an arbitrary, not necessarily conjugate, pair of planes

If the object goes to infinity $(s, y \to \infty)$, s' coincides with the posterior focal length f' of $\mathbb{S}$. Therefore, since $\theta = -y_1/s$, from the previous relation we have

$$f' = \frac{N'_n}{N_1} \frac{y'_n}{\theta}. \qquad (2.53)$$

- Finally, if π_0 and π'_0 represent an *arbitrary* and not necessarily conjugate pair of planes, from Figure 2.11 we can easily obtain

$$y_1 = -s\alpha_p + Y_p, \quad y'_n = -s'\alpha'_p + Y'_p,$$

so (2.52) can be written as follows:

$$N_1 Y_p \alpha_1 - N_1 Y_1 \alpha_p = N'_n Y'_p \alpha'_n - N'_n Y \alpha'_p = N'_n y'_n \alpha'_n, \qquad (2.54)$$

where $Y_1 = \alpha_1 s$ and $Y'_n = \alpha'_n s'$. This relation is called **Lagrange's invariant** and reduces to Helmholtz's invariant when $\pi_0 = \pi_e$ and $\pi'_0 = \pi'_e$.

2.6 Gaussian Analysis of Compound Systems

It has already been emphasized that the transformation (2.36)-(2.37) can be used to determine the image formed by an axially symmetric optical system $\mathbb{S}$ provided that y_1, α_1 are first-order quantities. In other words, the results of the above transformation are acceptable only for points next to the optical axis a and for rays that form small angles with a. However, the paraxial approximation is very important, since the paraxial behavior of an optical system usually represents its ideal behavior. For this reason, defects or the aberrations in $\mathbb{S}$ are defined as deviations from the paraxial response.

2.6. Gaussian Analysis of Compound Systems

In this section a complete Gaussian description of an optical system $\mathbb{S}$ is generated in detail using all of the considerations discussed in the above sections. The subject is presented a way that facilitates its implementation in a programming language.

- First, the matrices (2.28) and (2.29)

$$\mathbb{R}_i = \begin{pmatrix} 1 & 0 \\ -\dfrac{P_i}{N'_i} & \dfrac{N_i}{N'_i} \end{pmatrix}, \qquad (2.55)$$

$$\mathbb{D}_i = \begin{pmatrix} 1 & t_i \\ 0 & 1 \end{pmatrix}, \qquad (2.56)$$

must be written for each surface in the system. These describe, respectively, the refraction at the surface S_i and the behavior of meridional rays in the space between S_i and S_{i+1}.

- Consequently, the matrices

$$\mathbb{B}_i = \begin{cases} \mathbb{D}_i \mathbb{R}_i, & i = 1, \ldots, n-1, \\ \mathbb{R}_n, & i = n, \end{cases} \qquad (2.57)$$

fully describe the behavior of meridional rays from S_i to S_{i+1}.

- Then, the three matrices

$$\mathbb{M} = \mathbb{B}_n \cdots \mathbb{B}_1, \qquad (2.58)$$

$$\mathbb{M}^{(en)} = \mathbb{R}_k \mathbb{B}_{k-1} \cdots \mathbb{B}_1, \quad \mathbb{M}^{(ex)} = \mathbb{B}_n \cdots \mathbb{B}_{k+1}, \qquad (2.59)$$

must be evaluated. In (2.59), the index k, $0 \leq k \leq n$, relates to the last surface S_k before the stop. The matrix $\mathbb{M}$ describes the behavior of meridional rays throughout the system, whereas $\mathbb{M}^{(en)}$ and $\mathbb{M}^{(ex)}$ represent the actions of these rays at the surfaces before and after the stop, respectively. It is worth noting that

$$\begin{aligned} \mathbb{M}^{(en)} &= \mathbb{I}, \quad \text{if } k = 0, \\ \mathbb{M}^{(ex)} &= \mathbb{I}, \quad \text{if } k = n. \end{aligned} \qquad (2.60)$$

- The distances z_p and z'_p of the principal planes from the surfaces S_1 and S_n, respectively, are given by the relations (see (2.42))

$$z_p = \frac{M_{22} - 1}{M_{21}}, \quad z'_p = \frac{1 - M_{11}}{M_{21}}. \qquad (2.61)$$

- Let t_s be the distance of the stop from the surface S_k. Since the entrance pupil π_e is the image of the stop formed by the surfaces that precede it, we evaluate the distance w_e of the entrance pupil from the surface S_1 using the formula for conjugate points (2.39):

$$w_e = \frac{M_{22}^{(en)} t_s + M_{12}^{(en)}}{M_{21}^{(en)} t_s + M_{11}^{(en)}}. \qquad (2.62)$$

If $k = 0$ (the whole system $\mathbb{S}$ follows the stop), note that, due to (2.53), $\mathbb{M}^{(en)}$ is the identity matrix and the above formula gives $w_e = t_s$. Moreover, if the stop is situated after the system $\mathbb{S}$, then $\mathbb{M}^{(en)} = \mathbb{M}$.

- Similarly, the distance t_{ex} of the stop from the surface S_{k+1} is given by

$$t_{ex} = \begin{cases} -t_s & \text{if } k = 0, \\ -(t_k - t_s) & \text{if } 0 < k. \end{cases}$$

Then, the distance w'_e of the exit pupil from the last surface S_n is

$$w'_e = -\frac{M_{11}^{(ex)} t_{ex} - M_{12}^{(ex)}}{M_{21}^{(ex)} t_{ex} - M_{22}^{(ex)}}. \qquad (2.63)$$

- If r_s denotes the radius of the aperture stop, the radii r_{en} and r_{ex} of the entrance and exit pupils are given by

$$r_{en} = \frac{r_s}{M_{11}^{(en)} + M_{21}^{(en)} t_s},$$

$$r_{ex} = r_s (M_{11}^{(ex)} + M_{21}^{(ex)} w').$$

- The distance z'_n of the image from the last surface S_n of the whole system $\mathbb{S}$ is given by the usual formula for conjugate points:

$$z'_n = -\frac{M_{11} z_1 - M_{12}}{M_{21} z_1 - M_{22}}, \qquad (2.64)$$

where z_1 denotes the distance of the object from S_1. In particular, when $z_1 \to \infty$, we obtain the back focal length f'_b:

$$f'_b = -\frac{M_{11}}{M_{21}}, \qquad (2.65)$$

so that the posterior focal length f' is given by (see (2.44))

$$f' = f'_b - z'_{np}. \qquad (2.66)$$

2.6. Gaussian Analysis of Compound Systems

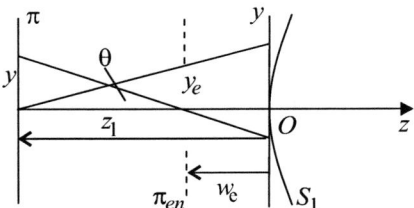

Fig. 2.12 View angle

- Finally, we need to evaluate the heights of the marginal ray and the principal ray on each surface S_i as well as the angles they form with the optical axis. Figure 2.12 depicts the marginal and principal rays.

It is easy to verify that the point at which these rays intercept the entrance pupil π_e as well as the angles that they form with the optical axis are given by the relations

$$\text{marginal ray} = \left(y_e, -\frac{y_e}{z_1 - w_e} \right), \qquad (2.67)$$

$$\text{principal ray} = \left(0, \frac{y}{z_1 - w_e} \right). \qquad (2.68)$$

In particular, if the object is situated at infinity, (2.67) and (2.68) become

$$\text{marginal ray} = (y_e, 0), \quad \text{principal ray} = (0, \theta). \qquad (2.69)$$

Consequently, the characteristics of these rays across the system are obtained by the relations

$$\begin{pmatrix} y'_i \\ \alpha'_i \end{pmatrix} = \mathbb{B}_i \cdots \mathbb{B}_1 \begin{pmatrix} 1 & -w_e \\ 0 & 1 \end{pmatrix} \begin{pmatrix} y \\ \alpha \end{pmatrix}, \qquad (2.70)$$

where $i = 1, \ldots, n-1$, and the matrix

$$\begin{pmatrix} 1 & -w_e \\ 0 & 1 \end{pmatrix}, \qquad (2.71)$$

describes the correspondence between the entrance pupil π_e and the tangent plane π_1^* at the vertex of S_1. In this formula, (y, α) is one of the rays (2.67) or (2.68) (or (2.69) if the object is at infinity). Finally, the equation

$$\begin{pmatrix} y'_n \\ \alpha'_n \end{pmatrix} = \begin{pmatrix} 1 & z'_n \\ 0 & 1 \end{pmatrix} \mathbb{M} \begin{pmatrix} 1 & -w_e \\ 0 & 1 \end{pmatrix} \begin{pmatrix} y \\ \alpha \end{pmatrix}, \qquad (2.72)$$

where z'_n is the distance of the image plane from the last surface S_n, supplies the characteristics of the marginal and principal rays that cross the image plane.

The notebook ***GaussianData*** supplies all of the Gaussian characteristics of a compound optical system.

2.7 A Graphical Method

When the characteristic points of $\mathbb{S}$ have been determined, it is possible to obtain the image of the object graphically using the simple procedure presented in this section. For the sake of simplicity, in what follows we refer to the most common case $N_1 = N'_n$, where the nodal and principal points coincide.

- Let y_1 be an object point and γ_0 be a ray parallel to a; in other words, a ray for which $\alpha_1 = 0$. Denote the point at which γ_0 (or its extension) meets the first principal plane π_p by y_p (see Figure 2.13).
  In the correspondence between the conjugate principal planes π_p and π'_p, the magnification is equal to 1. Consequently, at the exit of the whole optical system, the ray γ_0, or its extension, will meet the posterior principal plane π'_p at a point $y'_p = y_p$.

- The transformation (2.37) between the object plane and the posterior focal plane π'_f, when the condition (2.34) is taken into account, is described by the following matrix:

$$\begin{pmatrix} 1 & -\dfrac{M_{11}}{M_{21}} \\ 0 & 1 \end{pmatrix} \begin{pmatrix} M_{11} & M_{12} \\ M_{21} & M_{22} \end{pmatrix} \begin{pmatrix} 1 & -z_1 \\ 0 & 1 \end{pmatrix},$$

which reduces to the form

$$\begin{pmatrix} 0 & A \\ M_{21} & B \end{pmatrix},$$

where A and B have suitable expressions. Consequently, the correspondence between the object plane and the focal one becomes

$$y'_f = A\alpha_1,$$
$$\alpha'_f = M_{21} y_1 + B\alpha_1.$$

In particular, these relations imply that any ray that starts from the object point y_1 and is parallel to the optical axis a ($\alpha_1 = 0$) meets the focal plane on a.

2.8. Exercises

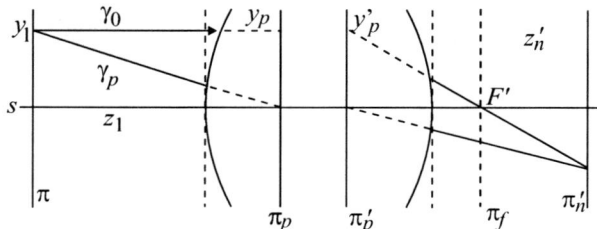

Fig. 2.13 Graphical construction of the image

- The ray γ_p, directed towards the anterior principal point (or node), has a corresponding parallel ray that intersects with the posterior principal point (or node).

Using these properties, the paraxial image can be obtained in the way illustrated in Figure 2.13.

When the refractive indices N and N' are different, it is sufficient to draw the latter conclusion for the nodal points.

2.8 Exercises

1. Verify that

$$\begin{pmatrix} \dfrac{N_i z_i'}{N_i' z_i} & 0 \\ -\dfrac{P_i}{N_i'} & \dfrac{z_i}{z_i'} \end{pmatrix} = \begin{pmatrix} 1 & z_i' \\ 0 & 1 \end{pmatrix} \begin{pmatrix} 1 & 0 \\ -\dfrac{P_i}{N_i'} & \dfrac{N_i}{N_i'} \end{pmatrix} \begin{pmatrix} 1 & -z_i \\ 0 & 1 \end{pmatrix}. \qquad (2.73)$$

The matrix composition on the right-hand side yields

$$\begin{pmatrix} 1 - \dfrac{z_i' P_i}{N_i'} & -z_i\left(1 - \dfrac{z_i' P_i}{N_i'}\right) + \dfrac{z_i' N_i}{N_i'} \\ -\dfrac{P_i}{N_i'} & \dfrac{z_i P_i}{N_i'} + \dfrac{N_i}{N_i'} \end{pmatrix}. \qquad (2.74)$$

On the other hand, from (2.7), it is easy to derive

$$\dfrac{N_i'}{z_i'} - \dfrac{N_i}{z_i} = P_i,$$

in other words
$$1 - \frac{P_i z_i'}{N_i'} = \frac{N_i z_i'}{N_i' z_i}.$$

It is now straightforward to show that the matrix (2.74) is equal to the left-hand side of (2.73).

2. In order to verify relation (2.46) and simplify the calculation, we suppose $n = 2$. We then obtain

$$\begin{pmatrix} \frac{N_2 z_2'}{N_2' z_2} & 0 \\ -\frac{P_2}{N_2'} & \frac{z_2}{z_2'} \end{pmatrix} \begin{pmatrix} \frac{N_1 z_1'}{N_1' z_1} & 0 \\ -\frac{P_1}{N_1'} & \frac{z_1}{z_1'} \end{pmatrix}$$

$$= \begin{pmatrix} M_1 M_2 & 0 \\ -M_1 \frac{P_2}{N_2'} & \frac{z_2 z_1}{z_2' z_1'} \end{pmatrix}.$$

On the other hand,

$$\frac{z_2 z_1}{z_2' z_1'} = \frac{N_2}{N_2'} \frac{N_2'}{N_2} \frac{z_2}{z_2'} \frac{z_1}{z_1'}$$
$$= \frac{N_1'}{N_2'} \frac{1}{M_2} \frac{z_1}{z_1'} = \frac{N_1}{N_2'} \frac{1}{M_2} \frac{N_1' z_1}{N_1 z_1'} = \frac{N_1}{N_2'} \frac{1}{M},$$

and the proof is complete.

3. Evaluate the Gaussian characteristics of a single spherical mirror S with a radius of curvature $R = -1000$ and an aperture stop that exists in the plane of the center of curvature.

The power of the mirror is

$$P = \frac{1}{500} = 0.002$$

and the resulting refractive matrix is

$$\mathbb{R} = \begin{pmatrix} 1 & 0 \\ 0.002 & -1 \end{pmatrix},$$

whereas the matrices $\mathbb{M}^{en}$ and $\mathbb{M}^{ex}$ (see Section 1.7) are

$$\mathbb{M}^{en} = \begin{pmatrix} 1 & 0 \\ 0 & 1 \end{pmatrix}, \quad \mathbb{M}^{ex} = \begin{pmatrix} 1 & 0 \\ 0.002 & -1 \end{pmatrix},$$

2.8. Exercises

such that the entrance and exit pupils are situated in the center of curvature. Finally, the matrix M for the whole system is

$$\mathbb{M} = \begin{pmatrix} 1 & 0 \\ 0.002 & -1 \end{pmatrix}.$$

Consequently, the distances of the principal planes from S (see (2.32) and (2.36)) are

$$z_{1p} = z_{2p} = 0$$

and the focal length (see (2.37)) is

$$f = -500.$$

4. Evaluate the Gaussian characteristics of a single plane-convex lens for which the anterior radius of curvature is 50, the posterior one is infinite, the thickness is 8 and the refractive index in green light is 1.518722. The aperture stop is −6, far from the lens.

The powers of the two surfaces are, respectively,

$$P_1 = 0.0103744, \quad P_2 = 0.$$

Consequently, the refractive matrices are

$$\mathbb{R}_1 = \begin{pmatrix} 1 & 0 \\ -0.0068 & 0.6584 \end{pmatrix}, \quad \mathbb{R}_2 = \begin{pmatrix} 1 & 0 \\ 0 & 1.518722 \end{pmatrix},$$

whereas the translation matrix $\mathbb{D}$ is

$$\mathbb{D} = \begin{pmatrix} 1 & 8 \\ 0 & 1 \end{pmatrix}.$$

Moreover, the matrices $\mathbb{M}^{en}$, $\mathbb{M}^{ex}$ and $\mathbb{M}$ are

$$\mathbb{M}^{en} = \begin{pmatrix} 1 & 0 \\ 0 & 1 \end{pmatrix}, \quad \mathbb{M}^{ex} = \begin{pmatrix} 0.9453 & 5.2675 \\ -0.0103 & 1 \end{pmatrix},$$

$$\mathbb{M} = \begin{pmatrix} 0.9453 & 5.2675 \\ -0.0103 & 1 \end{pmatrix}.$$

From the previous matrices, we can derive

$$z_{1p} = 0, \quad z_{2p} = -5.267,$$

$$f = 96.39.$$

Evaluate the distance of the image z'_2 when $z_1 = -200$.

5. Evaluate the Gaussian characteristics of the following optical system: the radii are $50, \infty, \infty, -50$; the distances are $8, 12, 8$; the refractive indices are $1, 1.518722, 1, 1.518722$; the aperture stop is situated between the second and third surfaces, and its distance from the second surface is 6.

We have

$$P_1 = 0.0104, \quad P_2 = P_3 = 0, \quad P_4 = 0.0104;$$

$$\mathbb{R}_1 = \begin{pmatrix} 1 & 0 \\ -0.0068 & 0.6584 \end{pmatrix}, \quad \mathbb{R}_2 = \begin{pmatrix} 1 & 0 \\ 0 & 1.5187 \end{pmatrix},$$

$$\mathbb{R}_3 = \begin{pmatrix} 1 & 0 \\ 0 & 0.6584 \end{pmatrix}, \quad \mathbb{R}_4 = \begin{pmatrix} 1 & 0 \\ -0.0103 & 1.5187 \end{pmatrix},$$

$$\mathbb{D}_1 = \mathbb{D}_3 = \begin{pmatrix} 1 & 8 \\ 0 & 1 \end{pmatrix}, \quad \mathbb{D}_2 = \begin{pmatrix} 1 & 12 \\ 0 & 1 \end{pmatrix}.$$

Consequently,

$$\mathbb{M}^{en} = \begin{pmatrix} 0.945 & 5.267 \\ -0.0103 & 1 \end{pmatrix}, \quad \mathbb{M}^{ex} = \begin{pmatrix} 1 & 5.267 \\ -0.0103 & 0.9453 \end{pmatrix},$$

$$\mathbb{M} = \begin{pmatrix} 0.762 & 22.53 \\ -0.018 & 0.766 \end{pmatrix};$$

$f = 54.57$, and the distance w_e of the entrance pupil from the first surface and the distance w'_e of the exit pupil from the last surface are

$$w_e = 12.76, \quad w'_e = 0.727.$$

6. In this exercise, another potential way of introducing the fundamental formulae of Gaussian optics is presented. Using the notation of Figure 2.14 and elementary results from Euclidean geometry, we can easily obtain

$$i = \alpha + \beta, \quad r = \beta - \alpha'. \tag{2.75}$$

In the paraxial approximation, the refraction law is

$$Ni = N'r, \tag{2.76}$$

where N and N' are the refractive indices of the media before and after surface S, respectively. Moreover, if R is the radius of S, we have

$$\alpha = -\frac{h}{z}, \quad \beta = \frac{h}{R}, \quad \alpha' = \frac{h}{z'}. \tag{2.77}$$

2.8. Exercises

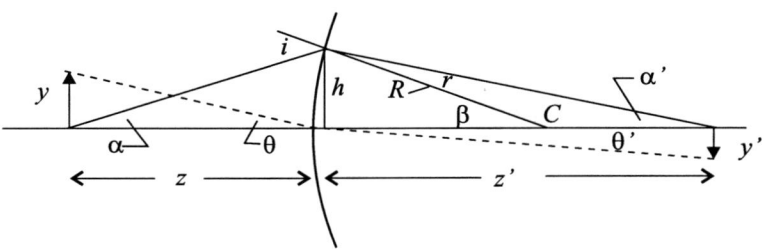

Fig. 2.14 Refraction law for an axial ray

Introducing (2.75) and (2.77) into (2.76), we obtain the Abbe invariant

$$\frac{N}{R} - \frac{N}{z} = \frac{N'}{R} - \frac{N'}{z'}.$$

Finally, by applying the refraction law to the principal ray, we deduce

$$N\theta = N'\theta',$$

which can also be written in the form

$$\frac{y'}{y} = \frac{z'N}{zN'}.$$

7. Using (2.40), (2.44), and (2.43), prove that (2.36) and (2.37) can also be written as follows:

$$y'_n = My_1, \qquad (2.78)$$

$$\alpha'_n = -\frac{1}{f'}y_1 + \frac{1}{M}\alpha_1. \qquad (2.79)$$

Chapter 3

Fermat's Principle and Third-Order Aberrations

3.1 Introduction

Starting from Fermat's principle, Hamilton introduced **characteristic functions** for analyzing the behavior of rays in a general optical system (see [17]). Subsequently, Seidel [18] and Schwarzschild [19] defined a new characteristic function termed **Schwarzschild's eikonal**. This is related to the Hamiltonian angle characteristic, but it depends on the use of suitable nondimensional variables. Using this function, the authors succeeded in developing a complete analysis, with evident geometrical meaning, of the third-order monochromatic aberrations of a compound optical system with a symmetry of revolution about the optical axis. Another approach to the optical aberration theory is based on the **aberration function** Φ_1, the determination of which again requires knowledge of Schwarzschild's eikonal (see, for instance, [9, 20]). All of these concepts will be discussed in Chapters 10 and 11, which are devoted to Hamiltonian optics.[1]

Whichever approach we choose to employ in order obtain to third-order aberration formulae, we must first understand the whole Hamiltonian formalism. In this chapter, a different approach is presented that is based only on Fermat's principle. It represents the natural generalization of the procedure we followed in order to introduce Gaussian optics. To be more specific, we introduce the classical aberration function Φ_1 via Fermat's principle and we show that knowledge of it allows us to determine aberrations in the optical system. Then we define **auxiliary optical paths**, each of which behaves in the optical system in a manner that is known a priori. Next, we introduce a new function Φ, which denotes the optical path length along an auxiliary optical path. Finally, we prove that this function, which

---
[1] For other, more recent, approaches to aberrations, see [21]–[24].

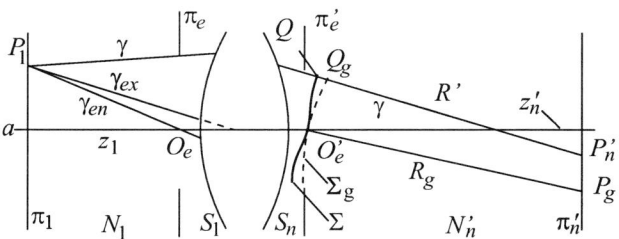

Fig. 3.1 Wavefront Σ and Gaussian reference sphere Σ_g

is easy to evaluate, is still an aberration function; in other words, it can be used to derive third-order monochromatic aberrations instead of Φ_1.

The notebook ***AberrationCoefficients*** supplies the third-order aberration coefficients of all of the optical systems we consider in this book. For more complex systems containing many optical elements with finite thicknesses, ***Mathematica***® may not be able to handle the calculations necessary to find the final third-order aberration coefficients. Even if it succeeds in this task, the expressions of these coefficients may be too complex to be useful.

3.2 The Aberration Function

Let $\mathbb{S}$ be an optical system with a symmetry of revolution about an optical axis a. We use the same notation we adopted in Chapter 2. In particular, z'_n denotes the distance of the image plane π'_n from the last surface S_n of $\mathbb{S}$, z_1 is the distance of the object plane π_1 from the first surface S_1 of $\mathbb{S}$, N'_n is the refractive index of the medium between S_n and S_{n+1}, and π_e and π'_e are the planes of the entrance and exit pupils of $\mathbb{S}$, respectively. Let $P_1 \equiv (0, y_1, z_1)$ be an object point and let $P_g \equiv (0, My_1, z'_n)$ be its Gaussian image, where M is the magnification of $\mathbb{S}$. Finally, we denote the point at which a ray γ originating from P_1 strikes the image plane π'_n by $P'_n \equiv (x'_n, y'_n, z'_n)$ (see Figure 3.1).

Let us consider the sphere Σ_g, which has its center at P_g and has a radius R_g equal to the distance between P_g and the center O'_e of the exit pupil. If w'_e is the distance of O'_e from the vertex of S_n, and we put

$$z_* = z'_n - w'_e, \qquad (3.1)$$

3.2. The Aberration Function

then the radius of Σ_g is
$$R_g^2 = z_*^2 + M^2 y_1^2, \tag{3.2}$$
and, in a reference frame that has its origin at O'_e, the equation for Σ_g assumes the form
$$f(X_g, Y_g, Z_g) \equiv X_g^2 + (Y_g - My_1)^2 + (Z_g - z_*)^2 - R_g^2 = 0, \tag{3.3}$$
where (X_g, Y_g, Z_g) are the coordinates of the moving point of Σ_g.

Let γ_{ex} be the ray from P_1 that crosses the exit pupil at $O'_e(0, 0, w'_e)$. In view of (1.5), we denote the optical path length along γ_{ex} from P_1 to O'_e by $V(y_1, z_1, w'_e)$. We now choose a point $Q(X, Y, Z)$ on any ray γ originating from P_1 such that the optical path length $V(y_1, z_1, X, Y, Z)$ from P_1 to Q along γ verifies the condition
$$V(y_1, z_1, X, Y, Z) = V(y_1, z_1, w'_e). \tag{3.4}$$

This equation defines a surface Σ (containing the point O'_e) that is the locus of the points $Q \in \gamma$ such that the optical path length along any γ from P_1 to Q has a value of $V(y_1, z_1, w'_e)$. Moreover, in view of (1.6), the ray γ is orthogonal to Σ at Q, since
$$\frac{\partial V}{\partial X}(y_1, z_1, X, Y, Z) = N'_n \alpha_x, \tag{3.5}$$
$$\frac{\partial V}{\partial Y}(y_1, z_1, X, Y, Z) = N'_n \alpha_y, \tag{3.6}$$
$$\frac{\partial V}{\partial Z}(y_1, z_1, X, Y, Z) = N'_n \alpha_z, \tag{3.7}$$
where α_x, α_y, and α_z are the director cosines of γ with respect to the coordinate axes.

Finally, if $Q_g(X_g, Y_g, Z_g)$ is the point of intersection of γ with the sphere Σ_g, then the function
$$\Phi_1 = V(y_1, z_1, X_g, Y_g, Z_g) - V(y_1, z_1, X, Y, Z) \tag{3.8}$$
is said to be the **aberration function** [9]. We note that, if $\Phi_1 = 0$, then the two surfaces Σ and Σ_g coincide and all of the rays—which for (3.5–3.7) are orthogonal to Σ—arrive at the same point P_g, so that there is no aberration.

Since the optical path length up to Σ is the same for all of the rays, γ in (3.8) can be replaced with another arbitrary but fixed ray. We choose the ray γ_{en}, which starts from P_1 and intercepts the center O_e of the entrance pupil. Since this ray is completely determined by these two points, the corresponding optical path length is a function of only y_1 and the distance

w_e of the entrance pupil from the first surface S_1 of $\mathbb{S}$. These remarks imply that the function Φ can also be written as follows:

$$\Phi_1 = V(y_1, z_1, X_g, Y_g, Z_g) - V(y_1, z_1, w_e), \qquad (3.9)$$

where $V(y_1, z_1, w_e)$ is the optical path length along γ_{en} up to Σ. Since Q_g is a point on the Gaussian sphere Σ_g, its coordinates verify Eq. 3.3; consequently, using Dini's theorem on implicit functions, we derive the result

$$\begin{aligned}\frac{\partial \Phi_1}{\partial X_g} &= \frac{\partial V}{\partial X_g} + \frac{\partial V}{\partial Z_g}\frac{\partial Z_g}{\partial X_g} \\ &= \frac{\partial V}{\partial X_g} - \frac{\partial V}{\partial Z_g}\frac{\frac{\partial f}{\partial X_g}}{\frac{\partial f}{\partial Z_g}} \\ &= \frac{\partial V}{\partial X_g} - \frac{\partial V}{\partial Z_g}\frac{X_g}{Z_g - z_*}. \end{aligned} \qquad (3.10)$$

On the other hand, in the reference frame with its origin at O'_e, we also have

$$\begin{aligned}\alpha_x &= \frac{x'_n - X_g}{R'}, \\ \alpha_y &= \frac{y'_n - Y_g}{R'}, \\ \alpha_z &= \frac{z_* - Z_g}{R'},\end{aligned} \qquad (3.11)$$

where R' is the distance between the points Q_g and P'_n (see Figure 3.1). Finally, collecting the results (3.5)–(3.7), (3.10), and (3.11) together, we have

$$\begin{aligned}\frac{\partial \Phi_1}{\partial X_g} &= N'_n \frac{x'_n}{R'} \\ \frac{\partial \Phi_1}{\partial Y_g} &= N'_n \frac{y'_n - My_1}{R'}.\end{aligned}$$

If we term $(\epsilon_x, \epsilon_y) = (x'_n, y'_n - My_1)$ the **aberration vector**, the above relations can then also be written as follows:

$$\epsilon_x = \frac{R'}{N'_n}\frac{\partial \Phi_1}{\partial X_g}, \qquad (3.12)$$

$$\epsilon_y = \frac{R'}{N'_n}\frac{\partial \Phi_1}{\partial Y_g}. \qquad (3.13)$$

Remark These relations, although exact, do not allow us to determine the aberrations of the system even when the aberration function is given,

3.2. The Aberration Function

since the quantity R' depends on the ray in an unknown way. To make them useful, we can expand R' in terms of known quantities so that we can use (3.12) and (3.13) to approximately determine the aberrations when Φ_1 is given.

Remark Due to the symmetry of revolution of $\mathbb{S}$ about the optical axis a, the aberration function $\Phi_1(y_1, X_g, Y_g)$ depends on the rotational invariants

$$\xi_g = y_1^2, \quad \eta_g = y_1 Y_g, \quad \zeta_g = X_g^2 + Y_g^2. \tag{3.14}$$

Consequently, the power expansion of Φ_1 in terms of these variables has the form

$$\Phi_1 = \Phi_1^{(1)} + \Phi_1^{(2)} + \Phi_1^{(3)} + \cdots + O_{inv}(n+1), \tag{3.15}$$

where $\Phi_1^{(n)}$ is a homogeneous polynomial in the variables ξ_g, η_g, and ζ_g, and $O_{inv}(n+1)$ denotes quantities of at least order $n+1$ in these invariants. Equivalently, the expansion contains only even powers of y_1, X_g, and Y_g. When we wish to relate the order of the neglected terms in (3.15) to these variables instead, then we write $O(2(n+1))$ rather than $O_{inv}(n+1)$.

Remark If the object and image planes are chosen according to Gaussian optics, the first term $\Phi_1^{(1)}$ of the expansion (3.15) vanishes since it corresponds to the Gaussian approximation in which there is no aberration. This implies that ϵ_x and ϵ_y depend on at least third-order powers of y_1, X_g, and Y_g.

In order to express R' in (3.12) and (3.13) in terms of known quantities, we first note that, in view of (3.2), (3.3), we have

$$\begin{aligned} R'^2 &= (x'_n - X_g)^2 + (y'_n - Y_g)^2 + (z_* - Z_g)^2 \\ &= x'^2_n - 2x'_n X_g + y'^2_n - 2y'_n Y_g + X_g^2 + Y_g^2 + (z_* - Z_g)^2 \\ &= x'^2_n - 2x'_n X_g + y'^2_n - 2y'_n Y_g + z_*^2 + 2My_1 Y_g. \end{aligned}$$

Taking into account the definitions $\epsilon_x = x'_n$, $\epsilon_y = y'_n - My_1$ and the last remark, we derive

$$\begin{aligned} R'^2 &= z_*^2 + M^2 y_1^2 + (\epsilon_x^2 + \epsilon_y^2) + 2My_1 \epsilon_y - 2(\epsilon_x X_g + \epsilon_y Y_g) \\ &= z_*^2 + M^2 y_1^2 + O(4). \end{aligned}$$

Finally, we have the approximate relation (see (2.5))

$$R' = z_* + \frac{M^2 y_1^2}{2z_*} + O(4), \tag{3.16}$$

which allows us to write (3.12) and (3.13) in the form

$$\epsilon_x = \frac{1}{N'_n}\left(z_* + \frac{M^2}{2z_*}y_1^2\right)\frac{\partial \Phi_1}{\partial X_g} + O(7), \tag{3.17}$$

$$\epsilon_y = \frac{1}{N'_n}\left(z_* + \frac{M^2}{2z_*}y_1^2\right)\frac{\partial \Phi_1}{\partial Y_g} + O(7). \tag{3.18}$$

The third-order quantities

$$\epsilon_x^{(3)} = \frac{z_*}{N'_n}\frac{\partial \Phi_1^{(2)}}{\partial X_g}, \tag{3.19}$$

$$\epsilon_y^{(3)} = \frac{z_*}{N'_n}\frac{\partial \Phi_1^{(2)}}{\partial Y_g}, \tag{3.20}$$

in the variables y_1, X_g, Y_g are called the **primary aberrations** or the **third-order aberrations**.

3.3 A New Aberration Function

Although the formulae (3.17) and (3.18) supply the aberrations of an optical system $\mathbb{S}$ to within seventh-order terms in the variables y_1, X_g, Y_g, they can only be used if we are able to determine

- The function $\Phi_1(X_g, Y_g)$

- The relations between the coordinates (X_g, Y_g) of the point $Q_g \in \Sigma_g$ and the coordinates of the entering ray γ, which are given by y_1 and the coordinates (x_e, y_e) of the point at which γ intersects with the entrance pupil.

Answering these questions is a very difficult task, even for the third-order aberrations, since we do not know the behavior of the rays. These problems will be discussed in the Hamiltonian framework in Chapter 8. In this section we define a new function $\Phi(y_1, x_e, y_e)$ that measures the optical path lengths of particular *known* paths that are not rays. Moreover, we prove that this new function $\Phi(y_1, x_e, y_e)$ can be used to evaluate third-order aberrations instead of Φ_1 [25].

First, we define these particular paths. Let $\mathbb{S}$ be an optical system with a symmetry of revolution about an axis a. If $P_1 \equiv (0, y_1, z_1)$ is an object point and $P_g \equiv (0, My_1, z'_n)$ is its Gaussian image, then the **stigmatic path** from P_1 to P_g associated with the ray γ is the broken line γ_s obtained with the following procedure. The path coincides with γ from P_1 up to the

3.3. A New Aberration Function

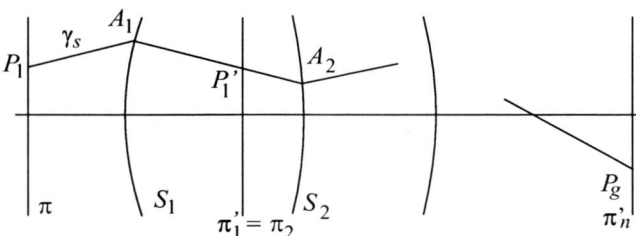

Fig. 3.2 Path of a stigmatic ray

point A_1, when γ intersects with the first surface S_1 of $\mathbb{S}$; then, denoting the magnification related to the planes z_1 and z'_1 by M_1, we consider the line from A_1 up to the Gaussian image $P'_1 \equiv (0, M_1 y_1, z'_1)$ formed by S_1 and the point A_2, when this line intersects with the second surface S_2 of $\mathbb{S}$. Proceeding in this way, we obtain a broken line $P_1 A_1 A_2 \cdots P_g$ that reaches the final point P_g (see Figure 3.2).

In conclusion, a unique stigmatic path γ_s can be associated with any ray γ; moreover, the behavior of γ_s is completely determined by Gaussian optics. In what follows, we denote the stigmatic path associated with the ray γ_{en} that enters the entrance pupil of $\mathbb{S}$ (see Figure 3.1) by γ_{en}.

In order to introduce the new aberration function Φ, we denote by $Q_s \equiv (X_s, Y_s, Z_s)$ the point on the Gaussian sphere Σ_g at which the stigmatic ray γ_s meets Σ_g (see Figure 3.3). Since the coordinates (X_s, Y_s, Z_s) of Q_s must verify Eq. 3.3 for Σ_g, we have $Z_s = Z_s(X_s, Y_s)$. Then, we consider the function

$$\Phi(y_1, X_s, Y_s) = L_s - L_{en}, \tag{3.21}$$

where L_s is the optical path length from the object point $(0, y_1, z_1)$ to Q_s along the stigmatic ray γ_s, and L_{en} is the optical path length along the stigmatic path γ_{en} from P to O'_e. By adding the radius R' to both L_s and L_{en}, we obtain the optical lengths Δ_s and Δ_{en} from P to P_g along the stigmatic paths γ_s and γ_{en}, respectively:

$$\Phi(y_1, X_s, Y_s) = \Delta_s - \Delta_{en}. \tag{3.22}$$

Remark In the absence of aberrations, the quantities Φ and Φ_1 vanish. In fact, if $\Phi_1 = 0$, the surfaces Σ and Σ_s coincide so that all of the rays orthogonal to Σ arrive at the Gaussian image P_n^* and vice versa. Analogously, if $\Phi = 0$, the optical lengths of all stigmatic paths are equal so that, according to Fermat's principle, they are rays that arrive at P_g.

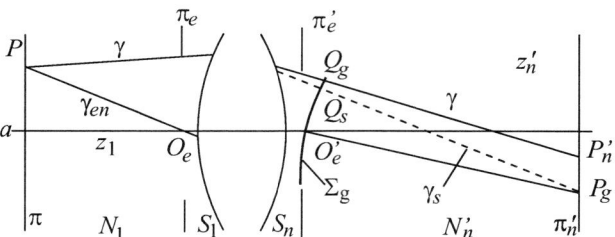

Fig. 3.3 New aberration function

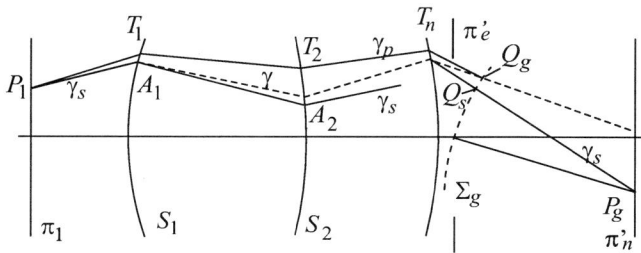

Fig. 3.4 Real and stigmatic rays

More generally, the following result holds.

Theorem 3.1
The functions Φ and Φ_1 verify the relations

$$\frac{\partial \Phi}{\partial X_g} = \frac{\partial \Phi_1}{\partial X_g} + O(5), \tag{3.23}$$

$$\frac{\partial \Phi}{\partial Y_g} = \frac{\partial \Phi_1}{\partial Y_g} + O(5), \tag{3.24}$$

where $O(5)$ denotes a fifth-order quantity in the variables X_g and Y_g.

PROOF Let γ be a ray originating from the object point $P_1 \equiv (0, y_1, z_1)$ and arriving at the point $Q_g \equiv (X_g, Y_g, Z_g)$ on the Gaussian sphere Σ_g. We denote the stigmatic path related to γ by γ_s and the point at which γ_s intersects Σ_g by $Q_s \equiv (X_s, Y_s, Z_s)$. Finally, let γ_p be any broken line starting from P_1, which has internal vertices $T_1, \ldots, T_n$ on the surfaces $S_1, \ldots, S_n$, and has a final point Q_g (see Figure 3.4).

3.3. A New Aberration Function

If $Z = f_i(X, Y)$ is the equation of S_i, then the optical path length along γ_p is given by the function

$$F(y_1, X_{T_1}, Y_{T_1}, \ldots, X_{T_n}, Y_{T_n}, X_{Q_g}, Y_{Q_g}, Z_{Q_g})$$
$$= F_1(y_1, X_{T_1}, Y_{T_1}, \ldots, X_{T_n}, Y_{T_n})$$
$$+ N'_n \sqrt{(X_g - X_{T_n})^2 + (Y_g - Y_{T_n})^2 + (Z_g - Z_{T_n})^2}, \quad (3.25)$$

where $F_1(y_1, X_{T_1}, Y_{T_1}, \ldots, X_{T_n}, Y_{T_n})$ denotes the optical path length along γ_p from P_1 to T_n. The variation of the optical path length F from γ_p to γ_s is

$$F(\gamma_s) - F(\gamma_p)$$
$$= \sum_{i=1}^{n} \left(\frac{\partial F_1}{\partial X_{T_i}} (X_{A_i} - X_{T_i}) + \frac{\partial F_1}{\partial Y_{T_i}} (Y_{A_i} - Y_{T_i}) \right)$$
$$+ \frac{N'_n}{l} \left((X_g - X_{T_n})(X_s - X_g) + (Y_g - Y_{T_n})(Y_s - Y_g) \right)$$
$$+ \frac{N'_n}{l} (Z_g - Z_{T_n})(Z_s - Z_g) + O(2), \quad (3.26)$$

where we have introduced the notation

$$l = \sqrt{(X_g - X_{T_n})^2 + (Y_g - Y_{T_n})^2 + (Z_g - Z_{T_n})^2},$$

and $O(2)$ is a second-order quantity with respect to the differences in the coordinates of the vertices of γ_p and γ_s.

On the other hand, if γ_p is the ray γ, then Fermat's principle implies that all of the derivatives of F_1 into (3.26) vanish and (1.6) holds. Therefore, (3.26) becomes (see (3.21)):

$$L_s - V(y_1, X, Y, Z(X, Y))$$
$$= \left(\frac{\partial V}{\partial X_g} \right)_{Q_g} (X_s - X_g) + \left(\frac{\partial V}{\partial Y_g} \right)_{Q_g} (Y_s - Y_g)$$
$$+ \left(\frac{\partial V}{\partial Z_g} \right)_{Q_g} (Z_s - Z_g) + O(2). \quad (3.27)$$

If the same procedure is applied to the pair γ_{en} and γ_{en}, then we obtain

$$L_{ens} - V(y_1, X_A, Y_A, Z(X_A, Y_A)) = \left(\frac{\partial V}{\partial X} \right)_{Q_A} (X_{As} - X_A) +$$
$$+ \left(\frac{\partial V}{\partial Y} \right)_{Q_A} (Y_{As} - Y_A) + \left(\frac{\partial V}{\partial Z} \right)_{Q_A} (Z_{As} - Z_A) + O(2), \quad (3.28)$$

where (X_A, Y_A, Z_A) and (X_{As}, Y_{As}, Z_{As}) denote the coordinates of the points at which γ_{en} and γ_{ens} intersect with Σ_g, respectively.

As we have already noted, the terms that appear in (3.28) depend only on the variable y_1, so (3.28) can also be written in the compact form

$$L_{ens} - \varphi(y_1) = \psi(y_1).$$

Finally, we have

$$(L_s - L_{ens}) - (V(y_1, X, Y, Z(X, Y)) - \varphi(y_1)) = \left(\frac{\partial V}{\partial X_g}\right)_{Q_g} (X_s - X_g)$$

$$+ \left(\frac{\partial V}{\partial Y_g}\right)_{Q_g} (Y_s - Y_g) + \left(\frac{\partial V}{\partial Z_g}\right)_{Q_g} (Z_s - Z_g) - \psi(y_1) + O(2).$$

Because of (3.8) and (3.21), this relation becomes

$$\Phi - \Phi_1 = \left(\frac{\partial V}{\partial X_g}\right)_{Q_g} (X_s - X_g) + \left(\frac{\partial V}{\partial Y_g}\right)_{Q_g} (Y_s - Y_g)$$

$$+ \left(\frac{\partial V}{\partial Z_g}\right)_{Q_g} (Z_s - Z_g) - \psi(y_1) + O(2). \quad (3.29)$$

Moreover, from (3.10) and (3.11) we derive

$$\frac{\partial V}{\partial X_g} = \frac{\partial \Phi_1}{\partial X_g} - \frac{\partial V}{\partial Z_g} \frac{\partial Z_g}{\partial X_g}, \quad (3.30)$$

$$\frac{\partial V}{\partial Y_g} = \frac{\partial \Phi_1}{\partial Y_g} - \frac{\partial V}{\partial Z_g} \frac{\partial Z_g}{\partial Y_g}, \quad (3.31)$$

$$\frac{\partial V}{\partial Z_g} = N'_n \alpha_z. \quad (3.32)$$

Since (X_s, Y_s, Z_s) and (X_g, Y_g, Z_g) belong to Σ_g, we get

$$Z_s - Z_g = \left(\frac{\partial Z_g}{\partial X_g}\right)_{Q_g} (X_s - X_g) + \left(\frac{\partial Z_g}{\partial Y_g}\right)_{Q_g} (Y_s - Y_g) + O_1(2),$$

where $O_1(2)$ is a second-order quantity with respect to the differences $X_s - X_g$ and $Y_s - Y_g$. Taking into account the above result, (3.29) assumes the form

$$\Phi - \Phi_1 = \left(\frac{\partial \Phi_1}{\partial X_g}\right)_{Q_g} (X_s - X_g) + \left(\frac{\partial \Phi_1}{\partial Y_g}\right)_{Q_g} (Y_s - Y_g)$$

$$- \psi(y_1) + O_2(2), \quad (3.33)$$

where $O_2(2) = N'_n \alpha_z O_1(2) + O(2)$.

Since $X_s = X_g$ and $Y_s = Y_g$ in the Gaussian approximation, we have

$$X_s = X_g + O(3), Y_s = Y_g + O(3) \quad (3.34)$$

so that (recalling that Φ_1 is a fourth-order function in the variables y_1, X_g, Y_g) we can write
$$\Phi - \Phi_1 = -\psi(y_1) + O(6),$$
where $O(6)$ denotes a sixth-order function of the variables X_g and Y_g, and we have proved the theorem. ∎

To conclude this section we note that the coordinates x'_e, y'_e in the exit pupil coincide with X_g, Y_g in the Gaussian approximation. Therefore, we have
$$X_g = M_e x_e + O(3), \quad Y_g = M_e x_e + O(3), \tag{3.35}$$
where M_e is the magnification related to the pupils, and the formulae (3.19) and (3.20), in view of (3.23) and (3.24), can also be written as follows:
$$\epsilon_x^{(3)} = \frac{z_*}{M_e N'_n} \frac{\partial \Phi^{(2)}}{\partial x_e}, \tag{3.36}$$
$$\epsilon_y^{(3)} = \frac{z_*}{M_e N'_n} \frac{\partial \Phi^{(2)}}{\partial y_e}. \tag{3.37}$$

In the following two sections the aberration function Φ will be evaluated for a compound optical system as a fourth-order function of the coordinates y_1, x_e, y_e, so that the third-order aberrations can be obtained from (3.36) and (3.37).

3.4 The Aberration Function Φ for a Single Surface

In this section we consider an optical system $\mathbb{S}$ containing a single surface S. To simplify the notation, we make $y_1 = y$, $z_1 = z$, $y'_1 = y'$, $z'_1 = z'$. The equation $Z = Z(X, Y)$ for S, up to fourth-order terms in the variables X and Y, is given by
$$Z = \frac{1}{2R}\eta + a_4 \eta^2 + O(3), \tag{3.38}$$
where $\eta = X^2 + Y^2$. In particular, if S is obtained by rotating a conic with a conic constant of K about a, we have (see Section 1.5)
$$a_4 = \frac{1+K}{8R^3}. \tag{3.39}$$

The following expansion:
$$\sqrt{a^2 + x + y^2} = a + \frac{x}{2a} - \frac{x^2}{8a^3} + \frac{y^2}{2a} + O(3) \tag{3.40}$$
allows us to prove the next theorem.

Theorem 3.2
Up to fourth-order terms in the coordinates (y, X, Y), the function $\Phi(y, X, Y)$ is given by the expression

$$\Phi = C_1 \eta^2 + \frac{1}{z} C_2 \eta \zeta + \frac{1}{z^2} C_3 \eta \xi + \frac{1}{z^2}(C_4 - C_3)\zeta^2 + \frac{1}{z^3} C_5 \xi \zeta + O(3), \quad (3.41)$$

where $\xi = y_1^2$, $\eta = X^2 + Y^2$, $\zeta = y_1 Y$, and

$$C_1 \equiv -\frac{N^2}{8}\left(\frac{1}{N'z'} - \frac{1}{Nz}\right)\left(\frac{1}{z} - \frac{1}{R}\right)^2 + \frac{N' - N}{8R^3} - (N' - N)a_4, \quad (3.42)$$

$$C_2 \equiv \frac{N^2}{2}\left(\frac{1}{N'z'} - \frac{1}{Nz}\right)\left(\frac{1}{z} - \frac{1}{R}\right), \quad (3.43)$$

$$C_3 \equiv \frac{N}{4}\left(\frac{N'^2 - N^2}{N'^2}\right)\left(\frac{1}{z} - \frac{1}{R}\right), \quad (3.44)$$

$$C_4 \equiv C_3 - \frac{N^2}{2}\left(\frac{1}{N'z'} - \frac{1}{Nz}\right), \quad (3.45)$$

$$C_5 \equiv -\frac{N}{2z^3} \frac{N'^2 - N^2}{N'^2}. \quad (3.46)$$

PROOF For a single surface S, the optical path difference (3.21) between the points $(y, 0, z)$ and $(My, 0, z')$ along the stigmatic path

$$\Phi = N\sqrt{X^2 + (Y - y)^2 + (Z - z)^2} + N'\sqrt{X^2 + (Y - My)^2 + (Z - z')^2},$$

when the variables ξ, η, and ζ are used, assumes the form

$$\Phi = N\sqrt{z^2 + \eta - 2\zeta + \xi - \frac{z}{R}\eta + \left(\frac{1}{4R^2} - 2a_4 z\right)\eta^2}$$

$$+ N'\sqrt{z'^2 + \eta - 2M\zeta + M^2\xi - \frac{z'}{R}\eta + \left(\frac{1}{4R^2} - 2a_4 z'\right)\eta^2}$$

$$- N\sqrt{\xi + z^2} - N'\sqrt{M^2\xi + z'^2}. \quad (3.47)$$

We identify the first-order quantity x appearing in (3.40) with the term $\eta - 2\zeta + \xi - \frac{z}{R}\eta$ of (3.47), the second-order quantity x^2 with the $(1/4R^2 - 2a_4 z)\eta^2$ terms, and we choose the values of z' and M in such a way as to eliminate the first-order terms of the expansion of Φ (see (2.7) and (2.8)).

3.4. The Aberration Function Φ for a Single Surface

In this way we have

$$\Phi = N\left(\frac{1}{8z^3}\left(\eta - 2\zeta + \xi - \frac{z}{R}\eta\right)^2 - \frac{1}{2z}\left(\frac{1}{4R^2} - 2a_4 z\right)\eta^2\right)$$
$$+ N'\left(-\frac{1}{8z'^3}\left(\eta - 2M\zeta + M^2\xi - \frac{z'}{R}\eta\right)^2 + \frac{1}{2z'}\left(\frac{1}{4R^2} - 2a_4 z'\right)\eta^2\right)$$
$$- \frac{N}{8z^3}\xi^2 + M^4 \frac{N'}{8z'^3}\xi^2. \tag{3.48}$$

Expanding the above expression, we obtain a quadratic with η^2, $\eta\zeta$, $\eta\xi$, ζ^2, and $\xi\zeta$ terms. The coefficient C_1 of η^2 is

$$C_1 = \frac{N}{8z}\left(\frac{1}{z^2} - \frac{2}{Rz}\right) + \frac{N'}{8z'}\left(-\frac{1}{z'^2} + \frac{2}{Rz'}\right) - (N' - N)a_4$$
$$= \frac{N}{8z}\left[\left(\frac{1}{z} - \frac{1}{R}\right)^2 - \frac{1}{R^2}\right] + \frac{N'}{8z'}\left[-\left(\frac{1}{z'} - \frac{1}{R}\right)^2 + \frac{1}{R^2}\right] - (N' - N)a_4$$
$$= -\frac{N^2}{8}\left(\frac{1}{N'z'} - \frac{1}{Nz}\right)\left(\frac{1}{z} - \frac{1}{R}\right)^2 + \frac{N' - N}{8R^3} - (N' - N)a_4, \tag{3.49}$$

provided that (2.7) is taken into account. Similarly, the coefficient C_2/z of $\eta\zeta$ is

$$\frac{C_2}{z} = -\frac{1}{2z^2}\left(\frac{N}{z} - \frac{N}{R}\right) + \frac{M}{2z'^2}\left(\frac{N'}{z'} - \frac{N'}{R}\right)$$
$$= \frac{N^2}{2z}\left(\frac{Mz}{Nz'^2} - \frac{1}{Nz}\right)\left(\frac{1}{z} - \frac{1}{R}\right)$$
$$= \frac{N^2}{2z}\left(\frac{1}{N'z'} - \frac{1}{Nz}\right)\left(\frac{1}{z} - \frac{1}{R}\right), \tag{3.50}$$

where (2.7) and (2.8) have been used. Moreover, the coefficient C_3/z^2 of $\eta\xi$ takes the form:

$$\frac{C_3}{z^2} = \frac{1}{4z^2}\left(\frac{N}{z} - \frac{N}{R}\right) - \frac{M^2}{4z'^2}\left(\frac{N'}{z'} - \frac{N'}{R}\right)$$
$$= \frac{1}{4}\left(\frac{1}{z^2} - \frac{M^2}{z'^2}\right)\left(\frac{N}{z} - \frac{N}{R}\right)$$
$$= \frac{N}{4z^2}\left(\frac{N'^2 - N^2}{N'^2}\right)\left(\frac{1}{z} - \frac{1}{R}\right), \tag{3.51}$$

whereas the coefficient $(C_4 - C_3)/z^2$ of ζ^2 becomes:

$$\frac{1}{z^2}(C_4 - C_3) = \frac{N}{2z^3} - \frac{M^2 N'}{2z'^3} = -\frac{N^2}{2z^2}\left(\frac{1}{N'z'} - \frac{1}{Nz}\right). \tag{3.52}$$

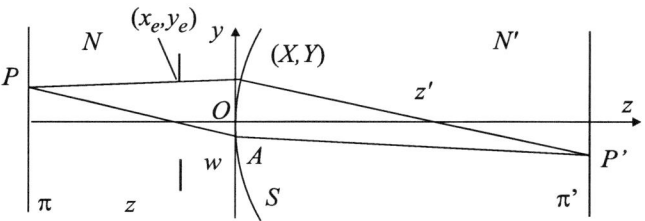

Fig. 3.5 Coordinates in the entrance pupil

Finally, the coefficient C_5/z^3 of $\xi\zeta$ is:

$$\frac{C_5}{z^3} = -\frac{N}{2z^3} + \frac{M^3 N'}{2z'^3} = -\frac{N}{2z^3}\frac{N'^2 - N^2}{N'^2}. \qquad (3.53)$$

In conclusion, the optical path difference Φ (assuming that the first-order part has been eliminated through appropriate selection of Gaussian conjugate planes and magnification) becomes

$$\Phi = C_1\eta^2 + \frac{1}{z}C_2\eta\zeta + \frac{1}{z^2}C_3\eta\xi + \frac{1}{z^2}(C_4 - C_3)\zeta^2 + \frac{1}{z^3}C_5\xi\zeta + O(3), \quad (3.54)$$

and we have proved the theorem. ∎

Now we must substitute the variables (X, Y) with the coordinates (x_e, y_e) of the point on the stigmatic path that intersects with the plane π_e of the entrance pupil (see Figure 3.5).

To this end, we consider the straight line

$$\begin{aligned} X_c &= Xt, \\ Y_c &= y + (Y - y)t, \\ Z_c &= z + (Z - z)t \end{aligned} \qquad (3.55)$$

between the points $(0, y, z) \in \pi$ and $(X, Y, Z) \in S_i$, where $Z = Z(X, Y)$ is Eq. 3.38 for S.

Denoting the distance of π_e from the vertex O of S by w, this straight line meets the plane π_e when

$$t = \frac{w - z}{Z - z} = \frac{z - w}{z}\left(1 + \frac{\eta}{2Rz_i}\right) + O(4). \qquad (3.56)$$

Upon substituting this value for the parameter into (3.55), the following functions are derived:

$$\begin{aligned} x_e &= \frac{z - w}{z}X + O(3), \\ y_e &= \frac{w}{z}y + \frac{z - w}{z}Y + O(3). \end{aligned} \qquad (3.57)$$

3.5. The Total Aberration Function for a Compound System

Inverting these functions, we obtain

$$X = cx_e + O(3),$$
$$Y = cy_e - dy + O(3), \qquad (3.58)$$

with

$$c = \frac{z}{z-w}, \quad d = \frac{w}{z-w}. \qquad (3.59)$$

In view of (3.58), the relation (3.54) then becomes

$$\Phi = C_1^* \eta^2 + \frac{c}{z} C_2^* \eta \zeta + \frac{c^2}{z^2} C_3^* \eta \xi + \frac{c^2}{z^2}(C_4^* - C_3^*) \zeta^2 + \frac{c^3}{z^3} C_5^* \xi \zeta + O(6), \qquad (3.60)$$

where the invariants ξ, η and ζ are now related to the variables y, x_e and y_e. Moreover,

$$C_1^* = c^4 C_1, \qquad (3.61)$$
$$C_2^* = c^2(C_2 - 4zdC_1), \qquad (3.62)$$
$$C_3^* = C_3 - zdC_2 + 2z^2 d^2 C_1, \qquad (3.63)$$
$$C_4^* = C_4 - 3zdC_2 + 6z^2 d^2 C_1, \qquad (3.64)$$
$$C_5^* = \frac{1}{c^2}(C_5 - 2zdC_4 + 3z^2 d^2 C_2 - 4z^3 d^3 C_1). \qquad (3.65)$$

In terms of (y, x_e, y_e), (3.60) becomes

$$\Phi = C_1^*(x_e^2 + y_e^2)^2 + \frac{c}{z} C_2^*(x_e^2 + y_e^2) y y_e + \frac{c^2}{z^2} C_3^*(x_e^2 + y_e^2) y^2$$
$$+ \frac{c^2}{z^2}(C_4^* - C_3^*) y^2 y_e^2 + \frac{c^3}{z^3} C_5^* y^3 y_e + O(6). \qquad (3.66)$$

The next sections are devoted to calculating the aberration function Φ of a compound system.

3.5 The Total Aberration Function for a Compound System

Let $\mathbb{S}$ be a compound optical system and π_e and π_e' be the planes of the entrance and exit pupils, respectively. Moreover, we put $\pi_{e,1} = \pi_e$ and denote the conjugate Gaussian plane of π_e with respect to the first surface S_1 of $\mathbb{S}$ by $\pi_{i,1}'$. Similarly, we put $\pi_{e,2} = \pi_{e,1}'$, and denote the conjugate Gaussian plane of $\pi_{e,2}$ with respect to the surface S_2 by $\pi_{e,2}'$, and so on.

It is evident that $\pi'_{e,n} = \pi'_e$. Finally, we denote the Gaussian magnification related to the pair of planes $\pi_{e,i}$ and $\pi'_{e,i}$ by $M_{e,i}$. Relation (3.66), referring to the surface S_i, then becomes

$$\Phi_i = C^*_{i,1}\eta_i^2 + \frac{c_i}{z_i}C^*_{i,2}\eta_i\zeta_i + \frac{c_i^2}{z_i^2}C^*_{i,3}\eta_i\xi_i$$
$$+ \frac{c_i^2}{z_i^2}(C^*_{i,4} - C^*_{i,3})\zeta_i^2 + \frac{c_i^3}{z_i^3}C^*_{i,5}\xi_i\zeta_i + O(6), \qquad (3.67)$$

where the coefficients are expressed by (3.61)–(3.65), referring to the surface S_i:

$$C^*_{i,1} = c_i^4 C_{i,1}, \qquad (3.68)$$
$$C^*_{i,2} = c_i^2(C_{i,2} - 4z_i d_i C_{i,1}), \qquad (3.69)$$
$$C^*_{i,3} = C_{i,3} - z_i d_i C_{i,2} + 2z_i^2 d_i^2 C_{i,1}, \qquad (3.70)$$
$$C^*_{i,4} = C_{i,4} - 3z_i d_i C_{i,2} + 6z_i^2 d_i^2 C_{i,1}, \qquad (3.71)$$
$$C^*_{i,5} = \frac{1}{c_i^2}(C_{i,5} - 2z_i d_i C_{i,4} + 3z_i^2 d_i^2 C_{i,2} - 4z_i^3 d_i^3 C_{i,1}), \qquad (3.72)$$

and the invariants ξ_i, η_i and ζ_i refer to the coordinates y_i, $x_{e,i}$ and $y_{e,i}$. Here, y_i denotes the ordinate of the Gaussian image on the plane π_i, and $x_{e,i}$, $y_{e,i}$ are the coordinates of the point of intersection of the stigmatic path with the plane $\pi_{e,i}$. Note that the orders of the neglected quantities refer to the coordinates y_i, $x_{e,i}$ and $y_{e,i}$, not to the corresponding invariants.

To express the total aberration function in terms of the invariants ξ_1, η_1 and ζ_1, related to the variables $y_1, x_{e,1} = x_e$ and $y_{e,1} = y_e$, we recall that

$$x'_{e,i} = M_{e,i} x_{e,i} + O(3), \qquad (3.73)$$
$$y'_{e,i} = M_{e,i} y_{e,i} + O(3), \qquad (3.74)$$
$$x_{e,i+1} = x'_{e,i}, \qquad (3.75)$$
$$y_{e,i+1} = y'_{e,i}. \qquad (3.76)$$

Then we have

$$\xi_i = y_i^2 = M_{1,i-1}^2 y_1 = M_{1,i-1}^2 \xi_1, \qquad (3.77)$$
$$\eta_i = x_{e,i}^2 + y_{e,i}^2 = M_{e,1,i-1}^2 \eta_1 + O(4), \qquad (3.78)$$
$$\zeta_i = y_i y_{e,i} = M_{1,i-1} M_{e,1,i-1} \zeta_1 + O(4) \qquad (3.79)$$

where $M_{e,1,i}$ is the Gaussian magnification between the planes $\pi_{e,1}$ and $\pi'_{e,i}$. Consequently, the total aberration function Φ becomes

$$\Phi = D_1 \eta_1^2 + \frac{c_1}{z_1} D_2 \eta_1 \zeta_1 + \frac{c_1^2}{z_1^2} D_3 \eta_1 \xi_1$$
$$+ \frac{c_1^2}{z_1^2}(D_4 - D_3)\zeta_1^2 + \frac{c_1^3}{z_1^3} D_5 \xi_1 \zeta_1 + O(6), \qquad (3.80)$$

3.5. The Total Aberration Function for a Compound System

where

$$D_1 = C^*_{1,1} + \sum_{i=2}^{n} M^4_{e,1,i-1} C^*_{i,1}, \qquad (3.81)$$

$$D_2 = C^*_{1,2} + \sum_{i=2}^{n} \frac{z_1}{z_i} \frac{c_i}{c_1} M^3_{e,1,i-1} M_{1,i-1} C^*_{i,2}, \qquad (3.82)$$

$$D_3 = C^*_{1,3} + \sum_{i=2}^{n} \left(\frac{z_1}{z_i}\frac{c_i}{c_1}\right)^2 M^2_{e,1,i-1} M^2_{1,i-1} C^*_{i,3}, \qquad (3.83)$$

$$D_4 = C^*_{1,4} + \sum_{i=2}^{n} \left(\frac{z_1}{z_i}\frac{c_i}{c_1}\right)^2 M^2_{e,1,i-1} M^2_{1,i-1} C^*_{i,4}, \qquad (3.84)$$

$$D_5 = C^*_{1,5} + \sum_{i=2}^{n} \left(\frac{z_1}{z_i}\frac{c_i}{c_1}\right)^3 M^2_{e,1,i-1} M^2_{1,i-1} C^*_{i,5}. \qquad (3.85)$$

If the object is located at infinity, ($z_1 \to \infty$), we have $y_1 \to \infty$ and the above formulae are meaningless. However, upon introducing the field angle

$$\theta = \frac{y_1}{z_1 - w} + O(3) = \frac{c_1}{z_1} y_1 + O(3) \qquad (3.86)$$

and noting that

$$\lim_{z_1 \to \infty} z_1 d_1 = w, \qquad \lim_{z_1 \to \infty} c_1 = 1, \qquad (3.87)$$

(3.80) assumes the following other form:

$$\Phi = D_1(x_e^2 + y_e^2)^2 + D_2(x_e^2 + y_e^2) y_e \theta$$
$$+ (D_3 x_e^2 + D_4 y_e^2) \theta^2 + D_5 y_e \theta^3 + O(6), \qquad (3.88)$$

where the notation $x_{e,1} = x_e$ and $y_{e,1} = y_e$ are again used.

It is possible to place the coefficients (3.81)-(3.85) in a simpler form by proving that

$$\frac{z_1}{z_i}\frac{c_i}{c_1} M_{e,1,i-1} M_{1,i-1} = \frac{N_1}{N_i}. \qquad (3.89)$$

In fact, with the help of Figure 3.6 and (3.59), it is straightforward to verify the following identities:

$$\frac{z_1}{z_i}\frac{c_i}{c_1} = \frac{z_1}{z_i}\frac{z_1 - w}{z_1}\frac{z_i}{z_i - w_i} = \frac{z_1 - w}{z_i - w_i} = \frac{-s_1}{z_i - w_i} = \frac{s_1}{s'_{i-1}}, \qquad (3.90)$$

since $s'_{i-1} = -(z_i - w_i)$. By resorting to (2.44), we have

$$\frac{z_1}{z_i}\frac{c_i}{c_1} = \frac{N_1}{N'_{i-1}} \frac{1}{M_{1,i-1} M_{e,1,i-1}}, \qquad (3.91)$$

and so (3.89) is proved when we recall that $N'_{i-1} = N_i$.

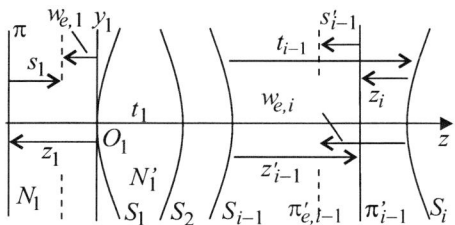

Fig. 3.6 Notation for a compound system

Finally, we arrive at the following simplified expressions for the coefficients D_i:

$$D_1 = C_{1,1}^* + \sum_{i=2}^{n} M_{e,1,i-1}^4 C_{i,1}^*, \qquad (3.92)$$

$$D_2 = C_{1,2}^* + N_1 \sum_{i=2}^{n} \frac{1}{N_i} M_{e,1,i-1}^2 C_{i,2}^*, \qquad (3.93)$$

$$D_3 = C_{1,3}^* + N_1^2 \sum_{i=2}^{n} \frac{1}{N_i^2} C_{i,3}^*, \qquad (3.94)$$

$$D_4 = C_{1,4}^* + N_1^2 \sum_{i=2}^{n} \frac{1}{N_i^2} C_{i,4}^*, \qquad (3.95)$$

$$D_5 = C_{1,5}^* + N_1^3 \sum_{i=2}^{n} \frac{1}{N_i^3 M_{e,1,i-1}^2} C_{i,5}^*. \qquad (3.96)$$

To evaluate the aberrations in the compound system, we insert (3.88) into (3.36) and (3.37) to obtain

$$\epsilon_x^{(3)} = \frac{z_n' - w_n'}{N_n' M_{e,1,n}} \left(4D_1(x_e^2 + y_e^2)x_e + 2D_2 x_e y_e \theta + 2D_3 x_e \theta^2 \right), \qquad (3.97)$$

$$\epsilon_y^{(3)} = \frac{z_n' - w_n'}{N_n' M_{e,1,n}} \left(4D_1(x_e^2 + y_e^2)y_e + 2D_2(x_e^2 + 3y_e^2)\theta + 2D_4 y_e \theta^2 + D_5 \theta^3 \right), \qquad (3.98)$$

where the coefficients $D_1, \ldots D_5$ are given by (3.92)–(3.96).

For the analysis presented in the next section, it is more convenient to write the above formulae by introducing polar coordinates (r_e, φ) for the entrance pupil.

In this coordinate system, we have

$$x_e = r_e \cos \varphi, \quad y_e = r_e \sin \varphi. \qquad (3.99)$$

3.6. Analysis of Third-Order Aberrations

Moreover, upon noting that

$$\cos^2 \varphi + 3\sin^2 \varphi = 3 - 2\cos^2 \varphi, \tag{3.100}$$

and taking into account the trigonometric identities

$$\cos 2\varphi = \cos^2 \varphi - \sin^2 \varphi = 2\cos^2 \varphi - 1, \tag{3.101}$$

we obtain

$$\cos^2 \varphi + 3\sin^2 \varphi = 2 - \cos 2\varphi. \tag{3.102}$$

It is now easy to change (3.97) and (3.98) into the following forms:

$$\epsilon_x^{(3)} = \mathbb{D}_1 r_e^3 \cos \varphi + \mathbb{D}_2 r_e^2 \theta \sin 2\varphi + \mathbb{D}_3 r_e \theta^2 \cos \varphi, \tag{3.103}$$

$$\epsilon_y^{(3)} = \mathbb{D}_1 r_e^3 \sin \varphi + \mathbb{D}_2 r_e^2 \theta (2 - \cos 2\varphi)$$
$$+ \mathbb{D}_4 r_e \theta^2 \sin \varphi + \mathbb{D}_5 \theta^3, \tag{3.104}$$

where

$$\mathbb{D}_1 = 4 \frac{z'_n - w'_n}{N'_n M_{e,1,n}} D_1, \tag{3.105}$$

$$\mathbb{D}_2 = \frac{z'_n - w'_n}{N'_n M_{e,1,n}} D_2, \tag{3.106}$$

$$\mathbb{D}_3 = 2 \frac{z'_n - w'_n}{N'_n M_{e,1,n}} D_3, \tag{3.107}$$

$$\mathbb{D}_4 = 2 \frac{z'_n - w'_n}{N'_n M_{e,1,n}} D_4, \tag{3.108}$$

$$\mathbb{D}_5 = \frac{z'_n - w'_n}{N'_n M_{e,1,n}} D_5. \tag{3.109}$$

3.6 Analysis of Third-Order Aberrations

In this section, the meaning of each term that appears in (3.65) as well as their combined effect on the final image are analyzed.

1. Spherical aberration. We consider rays originating from the axial object point. For all these rays, $\theta = 0$ and (3.103) and (3.104) supply the conditions

$$\epsilon_x^{(3)} = \mathbb{D}_1 r_e^3 \cos \varphi, \tag{3.110}$$

$$\epsilon_y^{(3)} = \mathbb{D}_1 r_e^3 \sin \varphi, \tag{3.111}$$

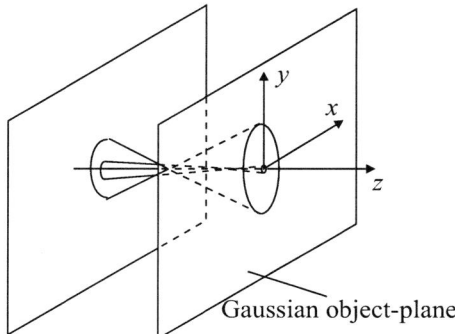

Fig. 3.7 Transverse spherical aberration

which, in turn, imply that

$$|\epsilon| = \sqrt{\epsilon_x^2 + \epsilon_y^2} = |\mathbb{D}|_1 r_e^3. \tag{3.112}$$

Equation 3.112 shows that all of the rays that originate from the axial object point and cross the entrance pupil along a circumference of radius r_e centered on the optical axis, instead of striking the *same* axial image point, spread along a circumference centered on the Gaussian image and with radius $\mathbb{D}_1 r_e^3$ (see Figure 3.7). The spherical aberration is *symmetric* around the optical axis. Moreover, let γ be a ray that intersects the entrance pupil at the point $(0, y_e)$, where $y_e > 0$. In this case, $\epsilon_x = 0$, $\epsilon_y = \mathbb{D}_1 y_e^3$; thus, if $\mathbb{D}_1 > 0$, ϵ_y has the same sign as y_e and γ meets the optical axis after the Gaussian plane. The opposite situation happens if $\mathbb{D}_1 > 0$.

We note that all of the rays that arise from the object point on the axis and intercept the entrance pupil along a circumference of radius r_e meet the optical axis at a same point. It can also be proved that the distance of this point from the Gaussian object plane is equal to

$$l(r_e) = \mathbb{D}_1 r_e^2. \tag{3.113}$$

2. **Coma.** In order to understand the contribution from the terms

$$\epsilon_x^{(3)} = \mathbb{D}_2 r_e^2 \theta \sin 2\varphi, \tag{3.114}$$
$$\epsilon_y^{(3)} = \mathbb{D}_2 r_e^2 \theta \left(2 - \cos 2\varphi\right), \tag{3.115}$$

to the aberrations, the set of rays from $(0, y)$ for which r_e is constant are considered. These rays intersect with the image plane in a circle of radius

$$R = \mathbb{D}_2 r_e^2 \theta, \tag{3.116}$$

3.6. Analysis of Third-Order Aberrations

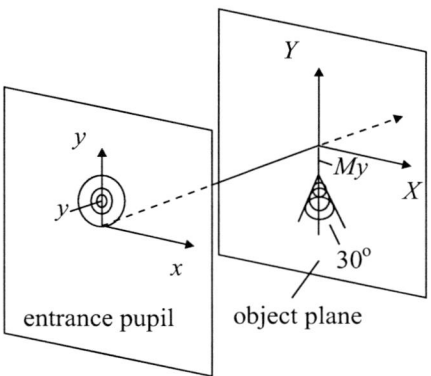

Fig. 3.8 Coma

that is centered at the point

$$(0, My + 2R). \tag{3.117}$$

The resulting image of a point in the object plane is given by the superposition of these circles upon varying r_e. If (X, Y) denotes a moving point in the object plane, the envelope of this one-parameter family Γ of circles can be obtained by eliminating the parameter R between the two equations

$$f(X, Y, R) = X^2 + (Y - 2R)^2 - R^2 = 0, \tag{3.118}$$
$$f_{,R}(X, Y, R) = -4(Y - 2R) - 2R = 0, \tag{3.119}$$

where (3.118) describes the family Γ and (3.119) is the derivative of the first equation with respect to the parameter R. A straightforward calculation leads to the following equation for the envelope:

$$X^2 - \frac{1}{3}Y^2 = 0, \tag{3.120}$$

which is therefore represented by two straight lines at an angle of 30° to the axis Oy.

This aberration, which is not symmetric, is called **coma** (see Figure 3.8).

3. Astigmatism and curvature of field. We now analyze the terms

$$\epsilon_x^{(3)} = \mathbb{D}_3 r_e \theta^2 \cos \varphi, \tag{3.121}$$
$$\epsilon_y^{(3)} = \mathbb{D}_4 r_e \theta^2 \sin \varphi, \tag{3.122}$$

for a given θ and r_e upon varying φ. Let y be the abscissa of the object point corresponding to the angle of view θ. We denote the set of all rays

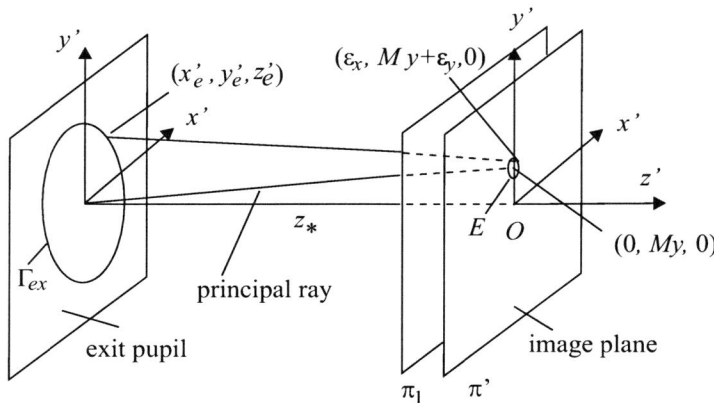

Fig. 3.9 Astigmatism and curvature of field

that originate from y and cross the exit pupil π'_e at a distance r'_e from the optical axis by Γ_{ex}. The rays of Γ_{ex} intercept an ellipse E with axes $\mathbb{D}_3 r_e \theta^2$ and $\mathbb{D}_4 r_e \theta^2$ that is centered at $(0, My)$ on the Gaussian image plane π'. Here M is the magnification of the optical system (see Figure 3.9).

To analyze this aberration we consider the curve of intersection of Γ_{ex} with another plane π_1 at a distance z_1 from π'. If (x'_1, y'_1) and (x', y') denote the coordinates in π_1 and π', respectively, then we can suppose that z_1 is of the order of y'^2.

The parametric equations for a ray from the point $(x'_e, y'_e) \in \Gamma_{ex}$ to the point $(\epsilon_x, My + \epsilon_y) \in \pi'$, when the origin of the axes is taken to be O, are:

$$X = \epsilon_x + (x'_e - \epsilon_x)t, \qquad (3.123)$$
$$Y = My + \epsilon_y + (y'_e - My - \epsilon_y)t, \qquad (3.124)$$
$$Z = -z_* t. \qquad (3.125)$$

The value of the parameter t when the ray intercepts the plane π_1 is

$$t = -\frac{z_1}{z_*}, \qquad (3.126)$$

so that the coordinates of the intersection point in the plane π_1 are

$$x_1 = \epsilon_x - x'_e \frac{z_1}{z_*}, \qquad (3.127)$$
$$y_1 = My + \epsilon_y - (y'_e - My)\frac{z_1}{z_*}. \qquad (3.128)$$

In writing the above equations we have taken into account the condition $z_1 \simeq M^2 y^2$ and noted that ϵ_x and ϵ_y are third-order quantities.

3.6. Analysis of Third-Order Aberrations

In view of (3.123) and (3.124), the relations (3.127) and (3.128) can also be written as follows:

$$x_1 = \left(\mathbb{D}_3 \theta^2 - M_{e,1,n} \frac{z_1}{z_*} \right) r_e \cos \varphi, \tag{3.129}$$

$$y_1 = My \left(1 + \frac{z_1}{z_*} \right) + \left(\mathbb{D}_4 \theta^2 - M_{e,1,n} \frac{z_1}{z_*} \right) r_e \sin \varphi. \tag{3.130}$$

where $M_{e,1,n}$ is the magnification between the pupils.

We can then conclude that the intersection of Γ_{ex} with the plane π_1 is again an ellipse with different axes and a different center. For a given θ, the axis along x' of this ellipse becomes zero when

$$z_1 = \frac{z_*}{M_{e,1,n}} \mathbb{D}_3 \, \theta^2 \quad \text{(meridional or tangential focus)}, \tag{3.131}$$

whereas the axis along y' vanishes when

$$z_2 = \frac{z_*}{M_{e,1,n}} \mathbb{D}_4 \, \theta^2 \quad \text{(primary or sagittal focus)}. \tag{3.132}$$

It is possible to change the above two relations into a form that allows us to interpret the geometrical meaning of the quantities $\mathbb{D}_3$ and $\mathbb{D}_4$. Recalling (2.51) and noting that $y' = \theta' z_*$, (3.131) and (3.132) then assume the following forms:

$$z_1 = \frac{z_* M_{e,1,n} N_n'^2}{N_1^2} \mathbb{D}_3 \theta'^2 = \frac{N_n'}{N_1^2} \frac{M_{e,1,n}}{z_*} \mathbb{D}_3 y'^2, \tag{3.133}$$

$$z_2 = \frac{z_* M_{e,1,n} N_n'^2}{N_1^2} \mathbb{D}_4 \theta'^2 = \frac{N_n'}{N_1^2} \frac{M_{e,1,n}}{z_*} \mathbb{D}_4 y'^2. \tag{3.134}$$

In conclusion, upon varying y', the loci of sagittal and tangential foci define two curves that, when rotated about the optical axis, define two spheres Σ_1 and Σ_2, which are called **sagittal and tangential or meridional focal surfaces**, respectively. The curvature radii of these surfaces are given by (see (3.107) and (3.108)):

$$\frac{1}{2R_s} = \frac{N_n'}{N_1^2} \frac{M_{e,1,n}}{z_*} \mathbb{D}_3 = 2\frac{1}{N_1^2} D_3, \tag{3.135}$$

$$\frac{1}{2R_t} = \frac{N_n'}{N_1^2} \frac{M_{e,1,n}}{z_*} \mathbb{D}_4 = 2\frac{1}{N_1^2} D_4. \tag{3.136}$$

The distance between Σ_1 and Σ_2

$$z_2 - z_1 = \frac{N_n'}{N_1^2} \frac{M_{e,1,n}}{z_*} (\mathbb{D}_4 - \mathbb{D}_3) y'^2 \tag{3.137}$$

is a measure of the **astigmatism** of the optical system.

4. Distortion. Using the last terms of (3.103) and (3.104)

$$\epsilon_x^{(3)} = 0, \tag{3.138}$$
$$\epsilon_y^{(3)} = \mathbb{D}_5 \theta^3, \tag{3.139}$$

we obtain

$$x' = 0, \tag{3.140}$$
$$y' = My + \mathbb{D}_5 \theta^3 = M\left(y + \frac{\mathbb{D}_5}{M}\theta^2\right). \tag{3.141}$$

Consequently, we conclude that the image point moves away from the axis if $\mathbb{D}_5/M > 0$ and towards the axis in the opposite case. These two cases are known as **pin-cushion and barrel distortion**, respectively.

3.7 Petzval's Theorem

We are now in the position to prove a very important theorem attributed to the Hungarian mathematician Petzval.

We will first prove the following relation:

$$D_4 - 3D_3 = -N_1^2 \sum_{i=1}^{n} \frac{1}{2R_i}\left(\frac{1}{N_i'} - \frac{1}{N_i}\right). \tag{3.142}$$

This relation is very interesting since it shows that, although D_4 and D_3 depend separately on the positions of the stop and the object, the difference $D_4 - 3D_3$ *depends only on the characteristics of the optical system*.

In order to prove (3.142), we start by noting that from (3.97) and (3.98) it follows that

$$C_{i,4} - 3C_{i,3} = -\frac{N_i^2}{2}\left(\frac{1}{N_i'z_i'} - \frac{1}{N_i z_i}\right)$$
$$- \frac{N_i}{2}\left(1 - \frac{N_i^2}{N_i'^2}\right)\left(\frac{1}{z_i} - \frac{1}{R_i}\right),$$

so that

$$C_{i,4} - 3C_{i,3} = -\frac{N_i^2}{2}\left(\frac{1}{N_i'z_i'} - \frac{1}{N_i z_i} + \left(\frac{1}{N_i'^2} - \frac{1}{N_i^2}\right)\left(\frac{N_i}{R_i} - \frac{N_i}{z_i}\right)\right).$$

Recalling Abbe's equation (2.7), here written in the equivalent form

$$\frac{1}{N_i'z_i'} = \frac{1}{N_i'R_i} - \frac{N_i}{N_i'^2 R_i} + \frac{N_i}{N_i'^2 z_i}, \tag{3.143}$$

3.7. Petzval's Theorem

the above condition becomes

$$C_{i,4} - 3C_{i,3} = -\frac{N_i^2}{2R_i}\left(\frac{1}{N_i'} - \frac{1}{N_i'}\right). \tag{3.144}$$

On the other hand, in view of (3.94) and (3.95), we also have

$$D_4 - 3D_3 = \left[C_{i,4} - 3C_{i,3} + \sum_{i=i}^{n}\frac{N_1^2}{N_i^2}(C_{i,4} - 3C_{i,3})\right], \tag{3.145}$$

and so we have proved (3.142).

Finally, (3.142), (3.135), and (3.136) imply that

$$\frac{1}{R_t} - \frac{3}{R_s} = 4\frac{1}{N_1^2}(D_4 - 3D_3). \tag{3.146}$$

Therefore, in view of (3.142), the following relation is obtained:

$$\frac{3}{R_s} - \frac{1}{R_t} = \frac{2}{R_p}, \tag{3.147}$$

where the quantity

$$\frac{1}{R_p} = \sum_{i=1}^{n}\frac{1}{R_i}\left(\frac{1}{N_i'} - \frac{1}{N_i}\right) \tag{3.148}$$

is called **Petzval's curvature**, whereas Eq. 3.147 expresses the contents of **Petzval's theorem**.

In particular, this implies that, in the absence of astigmatism (i.e., when $D_4 = D_3$) or equivalently when $1/R_s = 1/R_t$, the two focal surfaces Σ_1 and Σ_2 coincide with Petzval's surface of radius R_p. In this case, the condition that must hold if there is to be no field curvature is given by

$$\frac{1}{R_p} = 0, \tag{3.149}$$

which is independent of the positions of the object and stop.

In conclusion, all the above considerations lead us to define the following expression:

$$\frac{z_n' - w_n'}{N_n' M_{e,n}}(D_4 - 3D_3)\theta^2 r_e = -\frac{1}{2}\frac{z_n' - w_n'}{N_n'^2 M_{e,n}}\frac{N_1^2}{R_p}, \tag{3.150}$$

which is the **curvature of field**.

3.8 Aberration Formulae

In this section all of the fundamental aberration formulae are collected together for the reader's convenience.

Spherical aberration

$$\text{Spherical aberration} = 4\frac{z'_n - w'_n}{N'_n M_{e,n}} D_1 r_e^3, \qquad (3.151)$$

$$D_1 = C^*_{1,1} + \sum_{i=2}^{n} M^4_{e,i} C^*_{i,1},$$

where r_e is the radius of the entrance pupil.

Sagittal coma

$$\text{Sag. Coma} = -\frac{z'_n - w'_n}{N'_n M_{e,n}} D_2 \theta r_e^2, \qquad (3.152)$$

$$D_2 = C^*_{1,2} + N_1 \sum_{i=2}^{n} \frac{1}{N_i} M^2_{e,i-1} C^*_{i,2},$$

Astigmatism

$$\text{Astigmatism} = \frac{z'_n - w'_n}{N'_n M_{e,n}} (D_4 - D_3) \theta^2 r_e, \qquad (3.153)$$

$$D_3 = C^*_{1,3} + N_1^2 \sum_{i=2}^{n} \frac{1}{N_i^2} C^*_{i,3},$$

$$D_4 = C^*_{1,4} + N_1^2 \sum_{i=2}^{n} \frac{1}{N_i^2} C^*_{i,4}.$$

Curvature of field

$$\text{Curvature of field} = \frac{z'_n - w'_n}{N'_n M_{e,n}} (D_4 - 3D_3) \theta^2 r_e$$

$$= -\frac{1}{2} \frac{z'_n - w'_n}{N'^2_n M_{e,n}} \frac{N_1^2}{R_p} \theta^2 r_e, \qquad (3.154)$$

$$\frac{1}{R_p} = N'_n \sum_{i=1}^{n} \frac{1}{R_i} \left(\frac{1}{N'_i} - \frac{1}{N_i} \right) \qquad (3.155)$$

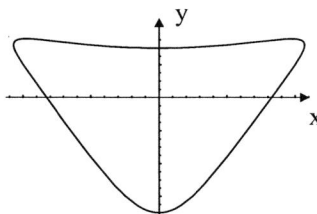

Fig. 3.10 A possible form of the curve Γ

Distortion

$$\text{Distortion} = \frac{z'_n - w'_n}{N'_n M_{e,n}} D_5 \theta^3, \tag{3.156}$$

$$D_5 = C^*_{1,5} + N_1^3 \sum_{i=2}^{n} \frac{1}{N_i^3 M_{e,i}^2} C^*_{i,5}.$$

3.9 Combined Effect of Third-Order Aberrations

In this section the combined effect of all third-order aberrations will be considered. To this end, we write formulae (3.103)–(3.104) as follows:

$$\epsilon_x^{(3)} = (\mathbb{D}_1 r_e^3 + \mathbb{D}_3 r_e \theta^2) \cos \varphi + \mathbb{D}_2 r_e^2 \theta \sin 2\varphi,$$
$$\epsilon_y = 2\mathbb{D}_2 r_e^2 \theta + \mathbb{D}_5 \theta^3 + (\mathbb{D}_1 r_e^3 + \mathbb{D}_4 r_e \theta^2) \sin \varphi - \mathbb{D}_2 r_e^2 \theta \cos 2\varphi.$$

and assume that r_e and y are constant. These equations can be regarded as the parametric equations of a curve in the Gaussian image plane π' with respect to the parameter φ. This curve can be constructed in the following way. We consider the ellipse E centered at the point

$$\left(0, 2\mathbb{D}_2 r_e^2 \theta + \mathbb{D}_5 \theta^3\right) \tag{3.157}$$

and with axes

$$\mathbb{D}_1 r_e^3 + \mathbb{D}_3 r_e \theta^2, \quad \mathbb{D}_1 r_e^3 + \mathbb{D}_4 r_e \theta^2. \tag{3.158}$$

Any point on this ellipse is then regarded as the center of a circle of radius $\mathbb{D}_2 r_e^2 \theta$. Finally, the resulting curve Γ behaves as shown in Figure 3.10.

It is clear that the form of Γ is strongly influenced by the characteristics of its component curves. For instance, Figure 3.11 shows two examples of Γ when the ellipse degenerates into a straight line.

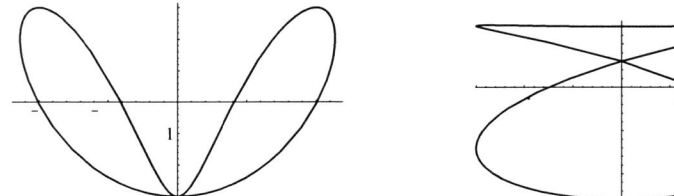

Fig. 3.11 Other possible forms of the curve Γ

Line	Wavelength (mm)	Element	Color
h	404.66	Hg	Violet
g	435.83	Hg	Violet
F'	479.99	Cd	Blue
F	486.13	H	Blue
e	546.07	Hg	Green
d	586.56	He	Yellow
C'	643.85	Cd	Red
C	656.27	H	Red
r	706.52	He	Deep red

Table 3.1 Wavelengths of different colors

3.10 Primary or Axial Chromatic Aberrations

In the previous two chapters, the refractive indices were considered to be characteristic constants of the media from which the lenses were made. However, it is well known that all substances exhibit wavelength-dependent refractive indices. This means that all of the results obtained so far only apply to a *fixed* wavelength. For this reason, the aberrations studied in the above sections are called monochromatic third-order aberrations.

A manufacturer of optical glasses will list the refractive indices at various standard wavelengths for each type of glass in its catalog (see Table 3.1).

The Abbe numbers

$$V_d = \frac{N_d - 1}{N_F - N_C}, \quad V_e = \frac{N_e - 1}{N_{F'} - N_{C'}}, \tag{3.159}$$

3.10. Primary or Axial Chromatic Aberrations

where N_λ denotes the refractive index for the wavelength λ, are a measure of the dispersion power of the glass under consideration.

The refractive index at the d line for real optical glasses varies from 1.44 to 1.96, whereas the Abbe numbers vary from 20 to 90. We note that a smaller Abbe number implies a higher dispersion power for the glass.

We begin our analysis of chromatic aberrations by studying the variations in the Gaussian characteristics of an optical system. More precisely, we evaluate *the change in the Gaussian position of the object and the change in the Gaussian magnification on varying the wavelength of the light.* The former change is termed the **axial chromatic aberration**, whereas the latter change is called the **chromatic magnification or lateral color aberration** (see also Section 1.4). In what follows, the notation Δa is introduced in order to denote the first-order variation in the quantity a, corresponding to a change in the refractive index ΔN.

For the first aberration, we begin by writing Abbe's invariant (2.7) in the equivalent form

$$\frac{N_i'}{z_i'} - \frac{N_i}{z_i} = \frac{N_i' - N_i}{R_i}. \tag{3.160}$$

Differentiating this relation, we derive the equation

$$-N_i'\frac{\Delta z_i'}{z_i'^2} + N_i\frac{\Delta z_i}{z_i^2} + \frac{\Delta N_i'}{z_i'} - \frac{\Delta N_i}{z_i} = \frac{\Delta N_i' - \Delta N_i}{R_i}, \tag{3.161}$$

which can be placed in the form

$$N_i'\frac{\Delta z_i'}{z_i'^2} - N_i\frac{\Delta z_i}{z_i^2} =$$
$$-\frac{\Delta N_i'}{N_i'}\left(\frac{N_i'}{R_i} - \frac{N_i'}{z_i'}\right) + \frac{\Delta N_i}{N_i}\left(\frac{N_i}{R_i} - \frac{N_i}{z_i}\right), \tag{3.162}$$

in other words

$$N_i'\frac{\Delta z_i'}{z_i'^2} - N_i\frac{\Delta z_i}{z_i^2} = -\left(\frac{\Delta N_i'}{N_i'} - \frac{\Delta N_i}{N_i}\right)Q_i, \tag{3.163}$$

where Q_i is Abbe's invariant for the surface S_i. We must now take into account the recursive relations

$$z_{i+1} = z_i' - t_i, \quad \Delta z_{i+1} = \Delta z_i', \quad N_{i+1} = N_i', \tag{3.164}$$

as well as the condition

$$\Delta z_1 = 0, \tag{3.165}$$

due to the fact that the position of the object is given.

With these results in mind, adding together all of the equations (3.163) obtained by allowing the index i to vary from 1 to n finally yields

$$\Delta z'_n = \frac{z'^2_n}{N'_n} \left[\sum_{i=1}^{n-1} N_{i+1} \Delta z_{i+1} \left(\frac{1}{(z'_i + t_i)^2} - \frac{1}{z'^2_i} \right) \right.$$
$$\left. - \sum_{i=1}^{n} \left(\frac{\Delta N'_i}{N'_i} - \frac{\Delta N_i}{N_i} \right) Q_i \right], \tag{3.166}$$

which expresses the changes in the image position as a function of the refractive indices and the characteristics of the optical system. In conclusion, the optical system will be free of axial chromatic aberration if $\Delta z'_n = 0$.

In particular, when lens thickness can be neglected, the above formula reduces to the following one:

$$\Delta z'_n = -\frac{z'^2_n}{N'_n} \sum_{i=1}^{n} \left(\frac{\Delta N'_i}{N'_i} - \frac{\Delta N_i}{N_i} \right) Q_i. \tag{3.167}$$

Finally, an approximate expression for (3.166) can be obtained using the power expansion

$$\frac{1}{(z'_i + t_i)^2} - \frac{1}{z'^2_i} = \frac{1}{z'^2_i} \left(-2 \frac{t_i}{z'_i} + 3 \frac{t_i^2}{z'^2_i} + \cdots \right). \tag{3.168}$$

It now remains to evaluate the change in the magnification M due to the wavelength-dependent nature of the refractive indices. Since $y_{i+1} = y'_i$, we have

$$M = \frac{y'_n}{y_1} = \frac{y'_n}{y_n} \frac{y'_{n-1}}{y_{n-1}} \cdots \frac{y'_1}{y_1}. \tag{3.169}$$

In view of (2.8), this relation becomes:

$$M = \frac{N_1}{N'_n} \frac{z'_1}{z_1} \frac{z'_2}{z_2} \cdots \frac{z'_n}{z_n}. \tag{3.170}$$

If f is any function, then the following formula holds:

$$\Delta (\log f) = \frac{\Delta f}{f}. \tag{3.171}$$

On the other hand, if $f = f_1 \cdots f_n$, then we have

$$\Delta(\log f) = \Delta(\log f_1 + \cdots \log f_n) = \frac{\Delta f_1}{f_1} + \cdots \frac{\Delta f_n}{f_n}, \tag{3.172}$$

so that

$$\frac{\Delta f}{f} = \frac{\Delta f_1}{f_1} + \cdots \frac{\Delta f_n}{f_n}. \tag{3.173}$$

By applying this relation to (3.170), we obtain the equation

$$\frac{\Delta M}{M} = \frac{\Delta N_1}{N_1} + N'_n \Delta\left(\frac{1}{N'_n}\right) + \sum_{i=1}^{n} \frac{\Delta z'_i}{z'_i} + \sum_{i=1}^{n} z_i \Delta\left(\frac{1}{z_i}\right), \qquad (3.174)$$

which can also be written

$$\frac{\Delta M}{M} = \frac{\Delta N_1}{N_1} - \frac{\Delta N'_n}{N'_n} + \sum_{i=1}^{n}\left(\frac{\Delta z'_i}{z'_i} - \frac{\Delta z_i}{z_i}\right). \qquad (3.175)$$

This formula supplies the chromatic variation in the magnification, since the changes Δz_i can be evaluated using (3.163) and (3.164).

3.11 Aplanatism and the Helmholtz Condition

So far we have considered third-order aberrations. In this section, we discuss the **sine** or **Helmholtz condition**, which allows us to control the coma in an optical system, correcting for the spherical aberration for **any** aperture angle. The relation that we are looking for is also called **Abbe's condition**, since Abbe understood its full meaning and applied it in an extensive way.

Let $\mathbb{S}$ be an optical system with a symmetry of revolution around the optical axis a. The object plane and the Gaussian image plane will be denoted by π and π', respectively, whereas O and O' are the points of intersection of a with π and π'. Finally, we assume that a stigmatic pair of points in π and π' does not exist, at least in the neighborhood of a. Using the notation shown in Figure 3.11 and recalling the results of Section 1.1, the optical path length $V(\mathbf{x}, \mathbf{x}')$ along the *unique* ray $\gamma(\mathbf{x}, \mathbf{x}')$ between two points $\mathbf{x} \in \pi$ and $\mathbf{x}' \in$ allows us to define the vectors tangent to $\gamma(\mathbf{x}, \mathbf{x}')$ at $\mathbf{x}$ and $\mathbf{x}'$ by employing relations (1.6):

$$\nabla_{\mathbf{x}} V(\mathbf{x}, \mathbf{x}') = -N(\mathbf{x})\mathbf{t}(\mathbf{x}), \quad \nabla_{\mathbf{x}'} V(\mathbf{x}, \mathbf{x}') = N(\mathbf{x}')\mathbf{t}(\mathbf{x}'). \qquad (3.176)$$

If $\mathbf{x}$ and $\mathbf{x}'$ are near to the optical axis a, we can expand the function $V(\mathbf{x}, \mathbf{x}')$ to within second-order terms as follows:

$$V(\mathbf{x}, \mathbf{x}') = V(\mathbf{O}, \mathbf{O}') + (\nabla V_{\mathbf{x}})_0 \cdot (\mathbf{x} - \mathbf{O}) + (\nabla V_{\mathbf{x}'})_0 \cdot (\mathbf{x}' - \mathbf{O}'), \qquad (3.177)$$

where the notation $(.)_0$ denotes that the gradients are evaluated at the points $\mathbf{O}$ and $\mathbf{O}'$. In view of (3.176), the above expansion can be written as

$$V(\mathbf{x}, \mathbf{x}') = V(\mathbf{O}, \mathbf{O}') - N\mathbf{t}_0 \cdot (\mathbf{x} - \mathbf{O}) + N'\mathbf{t}''_0 \cdot (\mathbf{x}' - \mathbf{O}'). \qquad (3.178)$$

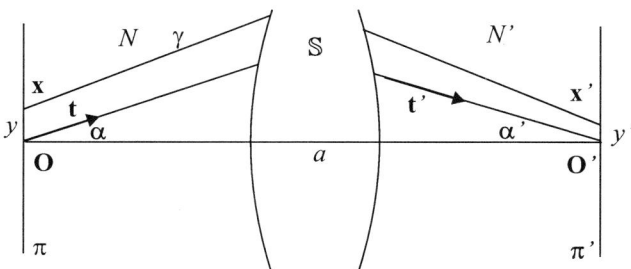

Fig. 3.12 Aplanatic (i.e., free from spherical aberration and coma) system

Since $\mathbf{x}$ and $\mathbf{x}'$ belong to the object and image planes respectively and the system exhibits symmetry of revolution, the scalar product $\mathbf{t}_O \cdot (\mathbf{x} - \mathbf{O})$ can always be written as $y \cos \beta$, where y is the coordinate of the object point with respect to the optical axis a and β denotes the angle between $\mathbf{t}$ and the axis Oy. Consequently, if α is the angle that $\mathbf{t}$ forms with a, we have $\mathbf{t}_0 \cdot (\mathbf{x} - \mathbf{O}) = y \sin \alpha$. Similar reasoning can also be applied to $\mathbf{t}'_{O'} \cdot (\mathbf{x}' - \mathbf{O}')$. Finally, (3.178) becomes

$$V(\mathbf{x}, \mathbf{x}') = V(\mathbf{O}, \mathbf{O}') - Ny \sin \alpha + N'y' \sin \alpha'. \tag{3.179}$$

In conclusion, let us suppose that

- $\mathbf{x}'$ is the Gaussian image of $\mathbf{x}$, so $y' = My$, where M is the magnification of K

- The sine condition
$$Ny \sin \alpha = N'y' \sin \alpha' \tag{3.180}$$
is satisfied

- The points $\mathbf{0}$ and $\mathbf{0}'$ on the axis are stigmatic ($V(\mathbf{0}, \mathbf{0}')$ does not depend on the ray between these two points); i.e., there is no spherical aberration for any α.

In this case, $V(\mathbf{x}, \mathbf{x}')$ does not depend on $\mathbf{x}$, $\mathbf{x}'$ either, and the system supplies a perfect image of any point $\mathbf{x}$ in the neighborhood of a belonging to π.

As an application of Abbe's condition, let us consider a spherical surface S that separates two media with refractive indices N and $N' < N$, respectively. We will now prove that, if $N' < N$ and $R < 0$,

- There are two points on the optical axis that are stigmatic
- The sine condition is satisfied for these two points.

3.11. Aplanatism and the Helmholtz Condition

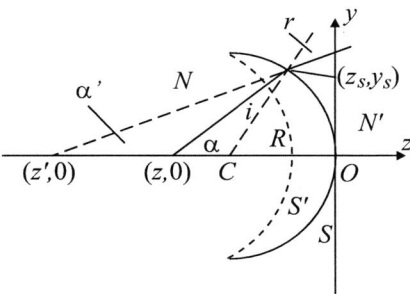

Fig. 3.13 Aplanatic meniscus

Using the notation shown in Figure 3.13, the condition that the optical path lengths of all meridional rays between the points $(z, 0)$ and $(z', 0)$ are the same yields

$$G = N\sqrt{y_s^2 + (z_s - z)^2} - N'\sqrt{y_s^2 + (z_s - z')^2} + Nz - N'z' = 0, \quad (3.181)$$

where R is the radius of the sphere S that is centered on C. The variable y_s can be eliminated by utilizing the sphere equation

$$y_s^2 = R^2 - (R - z_s)^2, \quad (3.182)$$

so that G reduces to a function of z_s. In other words, (3.181) is equivalent to requiring that

$$G'(z_s) = 0, \quad \forall |z_s| < |R|.$$

On the other hand, this condition can be written as

$$\frac{N(R - z)}{\sqrt{R^2 - (z_s - R)^2 + (z_s - z)^2}} - \frac{N'(R - z')}{\sqrt{R^2 - (z_s - R)^2 + (z_s - z')^2}} = 0,$$

and it is straightforward to verify that the previous condition is satisfied for any compatible value of z_s when

$$N'(z' - R)\sqrt{z^2 + 2Rz_s - 2zz_s} = N(z - R)\sqrt{z'^2 + 2Rz_s - 2z'z_s}.$$

Following simple but tedious calculations, this condition reduces to the following relation:

$$zz_s(R - z)(R - z')(N^2(R - z) - N'^2(R - z')) = 0,$$

which is satisfied for any value of z_s, and for negative values of z and z', when

$$z = R, \quad z' = R, \tag{3.183}$$

$$\frac{R-z}{R-z'} = \frac{N'^2}{N^2}. \tag{3.184}$$

It now remains to impose the sine condition on the solution $(3.184)_2$, since it is not valid for the simple case (3.183).

Applying the sine theorem to the triangles $(R,0), (z_s, y_s), (z, 0)$ and $(R,0), (z_s, y_s), (z', 0)$ (see Figure 3.13) yields the relations

$$-\frac{R}{\sin\alpha} = \frac{R-z}{\sin i}, \quad -\frac{R}{\sin\alpha'} = \frac{R-z'}{\sin r},$$

where i and r are the angles of incidence and refraction, respectively. Therefore, taking the refraction law

$$\frac{\sin r}{\sin i} = \frac{N}{N'},$$

into account, we have

$$\frac{\sin\alpha'}{\sin\alpha} = \frac{R-z}{R-z'}\frac{\sin r}{\sin i} = \frac{N}{N'}\frac{R-z}{R-z'}.$$

Finally, from (3.184) and the above relation, we obtain

$$\frac{\sin\alpha'}{\sin\alpha} = \frac{N}{N'}.$$

Consequently, we also have

$$N'y'\sin\alpha' = \frac{N'^2}{N}y'\sin\alpha = \frac{N'^2}{N^2}Ny\sin\alpha.$$

Recalling the magnification formula $y'/y = Nz'/N'z$, we can then derive the condition

$$N'y'\sin\alpha' = \frac{N'}{N}\frac{z'}{z}Ny\sin\alpha,$$

from which we obtain

$$\frac{N'z'}{Nz} = 1. \tag{3.185}$$

In conclusion, we obtain the values of z and z' from (3.184) and (3.185):

$$z = \frac{N+N'}{N}R, \quad z' = \frac{N'+N}{N'}R. \tag{3.186}$$

In particular, z, z', and R have the same sign.

3.12. Some Applications

We note that the object point $(z, 0)$ is located in the medium with a refractive index of N'. This condition can be achieved in two ways. First, a planar surface orthogonal to the axis Oz at a distance $d < |z + R|$ can be inserted, and the remaining space can then be filled with oil (with a refractive index of N'). Alternatively, a second spherical surface S' can be inserted (see Figure 3.13) that is centered on $(z, 0)$. This surface exerts no refractive effects on rays originating from this point.

3.12 Some Applications

In this section we will consider two simple applications of the third-order aberration formulae. Many other applications can be found in the notebook **TotalAberrations**.

Plate with plane parallel faces. Let K be a plate with plane parallel faces that has a refractive index N and a thickness d, and is surrounded by air (see Figure 3.14).

Applying (1.8) to the first surface S_1, we get

$$z'_1 = N z_1,$$

whereas we get the following for the second surface S_2:

$$z'_2 = \frac{z'_1 - d}{N} = z_1 - \frac{d}{N}.$$

We can conclude that the Gaussian image of the point Q, regardless of its position, is shifted along the straight line QQ' (which is orthogonal to S_1 and S_2) by an amount

$$\Delta = z'_2 + d - z_1 = d\left(1 - \frac{1}{N}\right). \tag{3.187}$$

Note that, since the refractive index N depends on the wavelength λ, the shift given by (3.187) varies with λ, meaning that K introduces an axial chromatism,

$$\Delta z' = d \frac{\Delta N}{N^2}.$$

We now wish to evaluate the spherical aberration and coma produced by K under the assumption that an aperture stop of radius r is located on the first face of K. Since the radii of curvature of the two faces of K are

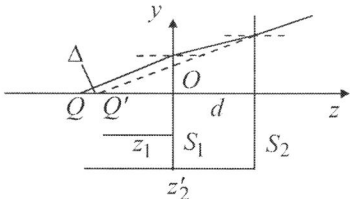

Fig. 3.14 Plane-parallel slab

equal to infinity, the formulae from Section 3.8 allow us to easily derive the following relations for the spherical aberration, coma and astigmatism, respectively:

$$\text{sph. ab.} = \frac{d(N^2-1)}{N^3 z^4} r^3, \tag{3.188}$$

$$\text{coma} = -\frac{d(N^2-1)}{N^3 z^3} r^2 y, \tag{3.189}$$

$$\text{sph. ab.} = \frac{d(N^2-1)}{N^3 z^2} r y_2, \tag{3.190}$$

$$\tag{3.191}$$

where y is the height of the object on the axis QQ'.

Aplanatic refractive surfaces. We will now search for a spherical refractive surface that is **aplanatic**; in other words, it has zero spherical aberration and coma for a fixed pair of conjugate points. This result, which is obtained using the third-order theory, coincides with the result derived using Abbe's condition in the previous section. Due to (3.34) and (3.35), the coefficients C_1 and C_2 vanish if

$$\frac{1}{N'z'} = \frac{1}{Nz},$$

so that

$$z' = \frac{N}{N'} z. \tag{3.192}$$

Substituting this result into (3.184), we obtain

$$R = \frac{N}{N' + N} z.$$

It is worth noting that these values are valid for any aperture stop (see (3.184)). By applying these formulae twice to the system shown in

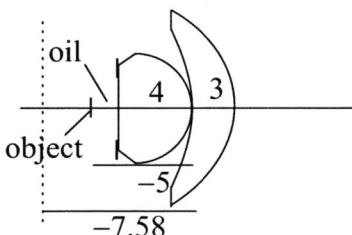

Fig. 3.15 Aplanatic pair of menisci

Figure 3.15, where the lenses are made from glass BK7 ($N = 1.5168$ in green light), we obtain the following data:

$$R_1 = \infty, \quad R_2 = -3.013, \quad d_1 = 5, \quad z_1' = -7.584$$
$$R_3 = -7.584, \quad R_4 = -6.378, \quad d_2 = 3, \quad z_2 = -16.05.$$

Evaluating the spherical aberration and coma for an aperture stop of 3 mm located at the first surface using the software package **OSLO**, we obtain

$$\text{sph. ab.} = 10^{-6}, \quad \text{coma} = -10^{-5}.$$

This type of aplanatic surface is used in oil immersion objectives and microscope condensers.

In contrast, the first surface of a dry microscope objective is centered on the object's axial point, implying that this type of objective has the further advantage of reducing the Petzval curvature of the image. Moreover, due to the presence of the cover glass for the specimen, the thickness of the first lens or the distance between the first and second lenses of the objective must be modified to introduce a residual spherical aberration in order to compensate for the spherical aberration of the cover glass.

3.13 Light Diffraction and the Airy Disk

All of the results obtained in this chapter have two fundamental limitations:

- Our analysis of aberrations was restricted to the third-order theory
- The geometric optics we used did not take into account the wave nature of light.

86   Chapter 3. Fermat's Principle and Third-Order Aberrations

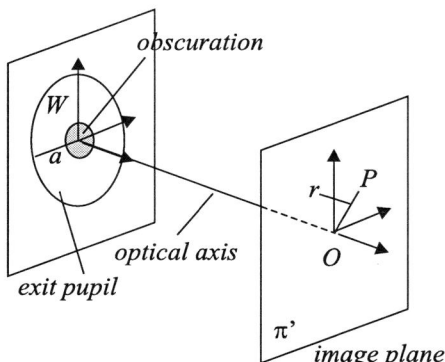

Fig. 3.16 Exit pupil with obscuration

It is possible to consider fifth-order aberrations, but the formulae required to do so are so cumbersome that they are of limited practical use. This implies that, after designing an optical system in which the third-order aberrations are eliminated, appropriate optimization techniques are then applied to optimize the performance of the system. These procedures, which must be used with great care, often lead to wrong results.

On the other hand, we cannot neglect the wave nature of light when designing an optical system, since it is this wave nature that determines how much the aberrations have to be corrected. In fact, due to its wave character, the propagation of light across an optical system produces (aside from reflections and refractions) diffraction phenomena. These phenomena lead to the fundamental result that the image of an object point, however perfect the optical system is, takes the form of a very small spot in which the intensity of light is distributed in a very complex way. The analysis of this problem is not easy and is carried out by applying suitable approximations that allow us to simplify the difficult calculations. Here, we limit ourselves to commenting on the final formulae and demonstrating some of their fundamental consequences. First, we suppose that the optical system $\mathbb{S}$ is ideal and has a symmetry of revolution about the optical axis, that the object is at infinity, and that the light is monochromatic with a wavelength of λ.

Under these conditions, the wavefront of the light W is a spherical surface that emerges from the exit pupil and is centered on the point O, as determined by the intersection of the Gaussian image plane π' with the optical axis (see Figure 3.16). We also suppose that the exit pupil has a radius a and an obscuration with a radius ϵa, and that it is located at a distance R from π'. The amplitude $U(r)$ of the wave at a point $P \in \pi'$ that is a distance r from O is a function of r. Moreover, we denote the intensity

3.13. Light Diffraction and the Airy Disk

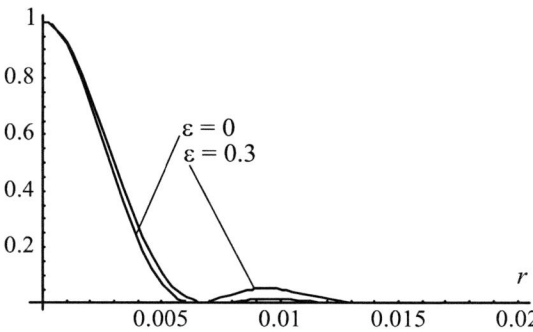

Fig. 3.17 Plot of the function $PSF(r)$

of the light at P by $I(r) = |U(r)|^2$, and the **point spread function** by $PSF(r) = I(r)/I(0)$. The following formula derived by Airy can be shown to be valid (see [9]):

$$PSF(r) = \frac{4}{(1-\epsilon^2)v^2}\left[J_1(v) - \epsilon J_1(\epsilon v)\right]^2, \qquad (3.193)$$

where

$$v = \frac{\pi}{\lambda}\frac{2a}{R}, \qquad (3.194)$$

$$J_1(v) = \sum_0^\infty \frac{(-1)^m}{m!\Gamma(m+2)}\left(\frac{v}{2}\right)^{2m+1} \qquad (3.195)$$

is a Bessel function of order 1, and

$$\Gamma(m+2) = \int_0^\infty e^{-x}x^{m+1}\,dx, \qquad (3.196)$$

is Euler's gamma function.

The behavior of the function $PSF(r)$ is shown in Figure 3.17 for $\lambda = 0.5875 \times (10)^{-3}$ mm. and $2a/R = 0.1$.

We can see from Figure 3.17 that the diffraction spot consists of a central disk—the **Airy disk**, which contains 84% of all of the light from the point source—surrounded by concentric rings that are alternately dark and bright. The light intensities of the bright concentric rings surrounding the Airy disk decrease according to the sequence 7%, 3%, and so on. It is apparent that the effect of the central obscuration is to transfer some of the energy from the first maximum to the other maxima.

In view of (3.193), the radii $r_1, r_2, \ldots$ of the bright rings can be derived by solving the equation

$$J_1(v) - \epsilon J_1(\epsilon v) = 0, \qquad (3.197)$$

ϵ	v_1	v_2	v_3
0.00	1.220π	2.233π	3.238π
0.10	1.205π	2.269π	3.182π
0.20	1.167π	2.357π	3.087π
0.33	1.098π	2.424π	3.137π
0.40	1.058π	2.388π	3.300π

Table 3.2 Obscuration factors and corresponding values of the roots v_1, v_2, and v_3

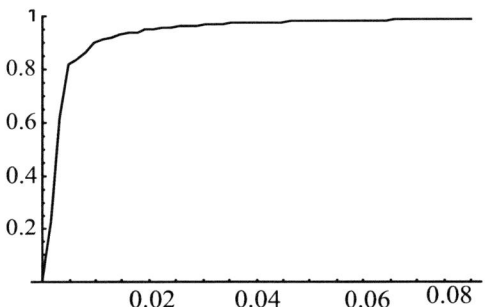

Fig. 3.18 Plot of the function $EE(r)$

and then using $SF2$. Table 3.2 lists the roots $v_1, v_2, \ldots$ of (3.197) for different values of obscuration ϵ.

In particular, if $R/2a$ is equal to the focal ratio A, then we find that the radius r_a of the first dark ring (i.e., of the **Airy ring**) is expressed by the formula

$$r_a = 1.22\lambda A. \tag{3.198}$$

It is also possible to determine the **encircled energy fraction** $EE(r)$, which gives the fraction of the total light energy of the image that is contained in a circle of radius r. It is possible to prove that the encircled energy fraction is expressed by the function

$$EE(r) = \frac{1}{(1-\epsilon^2)} \left[1 - J_0(v)^2 - J_1(v)^2 + \right. \tag{3.199}$$

$$\epsilon^2 \left(1 - J_0(\epsilon v)^2 - J_1(\epsilon v)^2\right) - \tag{3.200}$$

$$\left. 4\epsilon \int_0^v J_1(\epsilon x) \frac{J_1(x)}{x} dx \right], \tag{3.201}$$

where $J_0(v)$ is a Bessel function of order 0. Figure 3.18 plots this function.

3.13. Light Diffraction and the Airy Disk

All the above considerations refer to the case of a spherical wavefront centered on the image point O on the optical axis. In the presence of aberrations W is not spherical and the distribution of light is modified from the ideal case. Readers interested in the complex formulae that apply in this situation should refer to [36, 37].

We conclude with the following fundamental rule: *an optical system is perfect or diffraction limited if the aberration spot resides inside the Airy disk.*

Chapter 4

Newtonian and Cassegrain Telescopes

4.1 Newtonian Telescopes

The simplest and most famous type of telescope was invented by Newton. It consists of a parabolic primary mirror S_1 and a flat secondary mirror S_2 that is situated in front of the primary and directs the converging cone of light arriving from the primary out of the tube for visual or photographic applications. The aperture stop A coincides with the rim of S_1 (see Figure 4.1).

First, the image of the object, which is located at infinity, lies in the focal plane π'_f of the primary mirror. If R is the curvature radius of S_1, then the focal length $f = R/2$ is derived from Abbe's invariant (see (2.7)). Since the secondary mirror does not introduce aberrations, only the contributions from S_1 need to be evaluated.[1] To this end, we note that the refractive indices are $N = -N' = 1$, the conic constant K of S_1 is equal to -1, $z = \infty$, and $z' = R/2$. Moreover, the distances $w = w'$ are equal to zero, so that

$$M_e = 1. \quad (4.1)$$

Consequently, the following expressions for the total spherical aberration, coma and the astigmatism coefficients can be obtained from (3.41)–(3.46) and (3.61)–(3.65):

$$\begin{aligned} C_1 = C_1^* = 0, \quad C_2 = C_2^* = \frac{1}{R^2}, \\ C_3 = C_3^* = 0, \quad C_4 = C_4^* = \frac{1}{R}. \end{aligned} \quad (4.2)$$

---
[1] The aberration coefficients derived in this chapter can be evaluated using the notebook *AberrationCoefficients*.

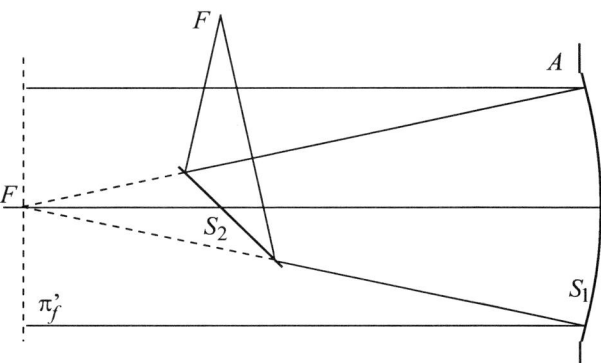

Fig. 4.1 Newtonian telescope

These relations show that there is no spherical aberration but there is coma, which is independent of the stop position. Put differently, the astigmatism coefficients yield the following relation:

$$D_4 - D_3 = C_4^* - C_3^* = \frac{1}{R}.$$

It should be noted that the astigmatism is not an important aberration for this combination. In fact, if we refer back to (3.97) and (3.98), we can obtain the following expressions for the sagittal coma and astigmatism:

$$\text{Sag. coma} = \frac{\theta d^2}{16f}, \quad \text{Astig.} = -\frac{1}{4}\theta^2 d, \qquad (4.3)$$

where d is the diameter of S_1 and θ is the field angle.

The ratio a between the astigmatism and the sagittal coma is

$$a = -\frac{4\theta f}{d}$$

and is much smaller than 1 under standard conditions. For instance, if the primary mirror has a diameter of 200 mm, a curvature radius of -1200 mm, and if the angle θ is $0.5° = 0.0087$ radians, we get $a = 0.10$. This means that coma is dominant—in other words, the limits of this optical combination are determined solely by this aberration.

- The size of the secondary mirror depends on the intended use of the instrument. If it is intended for visual use, the secondary must ensure that a region in the focal plane about 10 mm in size is illuminated without vignetting. On the other hand, it needs to be larger when the

4.1. Newtonian Telescopes

system is to be used for photographic purposes. However, whatever the intended use, it should not be made larger than necessary, as any obstruction of the primary degrades the quality of the final image. A drop in illumination of about 20%–30% at the corner of the photographic film is allowed. Moreover, the obstruction (i.e., the reduction in the amount of light entering the telescope) due to the secondary mirror must not be greater than 0.33.

- Let R and d denote the radius of curvature and the diameter of S_1, respectively. The focal length f of S_1 is then equal to $|R|/2$, $M_e = 1$, and the aperture ratio is

$$A = \frac{d}{f}. \tag{4.4}$$

- According to (3.198), the diameter of the Airy disk can be found using:

$$d_a = 2.44\lambda \frac{f}{d} = 2.44 \frac{\lambda}{A}, \tag{4.5}$$

since $w' = 0$ and $z' = f = |R|/2$. Consequently, if astigmatism is neglected, then the diameter of the geometrical spot due to coma is double that given by the first equation of (4.3), and the optical combination is diffraction limited if

$$\frac{\theta d^2}{8f} \leq 2.44\lambda \frac{f}{d},$$

leading to the following condition for the field angle:

$$\theta \simeq 2.44\lambda \frac{8f^2}{d^3}. \tag{4.6}$$

Consequently, the useful field diameter σ in the focal plane is

$$\sigma = 2.44\lambda \frac{16f^3}{d^3}, \tag{4.7}$$

provided that the instrument is intended for visual use.

However, if it is intended for photographic use, we obtain

$$2\frac{\theta d^2}{16f} \leq 0.025, \tag{4.8}$$

where 0.025 mm is the photographic grain diameter, and the useful photographic field is then

$$\sigma \leq 0.4 \frac{f^2}{d^2}. \tag{4.9}$$

Focal length	$f = \dfrac{|R|}{2}$
Diameter of useful visual field	$\sigma \leq 2.44\lambda\dfrac{16f^3}{d^3}$
Diameter of useful photographic field	$\sigma \leq 0.4\dfrac{f^2}{d^2}$

Table 4.1 Visual and photographic linear fields of a parabolic mirror

We conclude the analysis of this optical combination by making the following remark. In order to intercept all of the light needed for visual or photographic use, the boundary of the plane mirror must have an elliptical profile. Moreover, it is clear from Figure 4.1 that its center is shifted vertically with respect to the optical axis.

All the results from this section are summarized in Table 4.1.

4.2 Cassegrain Telescopes

In this section, the family of Cassegrain telescopes is analyzed. All Cassegrain telescopes consist of a concave primary mirror S_1 with a small convex secondary mirror S_2 situated between the primary focus F_1 and S_1. The light that arrives at S_1 is reflected back towards S_2. Then, after another reflection, it arrives at the eyepiece or the film placed across a hole centered on the vertex of S_1 (see Figure 4.2). The aperture stop is located at the primary mirror.

Usually, the design data are:

- The diameter d_1 of S_1

- The focal length $f_1 > 0$ of S_1 or, equivalently, its radius of curvature $R = -2f_1$

- The total focal length $f > 0$ of the combination

- The back distance $e > 0$ from S_1 to the posterior focal point.

We now show that all of the remaining data (i.e., d_2, R_2, and d) are determined by simple geometrical and Gaussian relations. On the other hand, since the imaging process results only from reflections, there is no chromatic aberration. The only way to control the third-order aberrations is

4.2. Cassegrain Telescopes

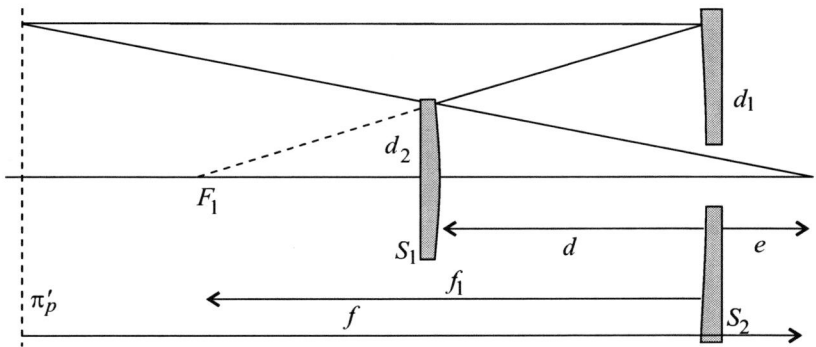

Fig. 4.2 Cassegrain telescope

to select the aspheric constants K_1 and K_2 of S_1 and S_2 wisely. In particular, this means that the most we can hope to achieve is to eliminate the spherical aberrations and coma.

Starting from the assigned data, the following nondimensional quantities can be evaluated:

$$\alpha = \frac{e}{f_1} > 0, \quad M = \frac{f}{f_1}. \tag{4.10}$$

The problem of determining d_2, R_2, and d is then equivalent to calculating the new nondimensional quantities

$$c = \frac{d_2}{d_1}, \quad \gamma = \frac{R_2}{f_1}, \quad \beta = -\frac{d}{f_1}, \tag{4.11}$$

where

$$c > 0, \quad \gamma < 0, \quad 0 < \beta < 1.$$

First, using Figure 4.2 and the third equation of (4.11), we derive the condition

$$c = \frac{d_2}{d_1} = \frac{f_1 + d}{f_1},$$

in other words

$$c = 1 - \beta. \tag{4.12}$$

From Figure 4.2 and the second equation of (4.10), we also have

$$c = \frac{d_2}{d_1} = \frac{-d + e}{f} = \frac{-d + e}{f_1} \frac{f_1}{f} = \frac{\alpha + \beta}{M}.$$

Therefore, taking into account (4.12), we obtain

$$\beta = \frac{M - \alpha}{M + 1}. \tag{4.13}$$

Finally, in order to find γ, we note that

$$z_2 = -f_1 - d, \quad z_2' = -d + e,$$
$$N_2 = -1, \quad N_2' = 1.$$

On the other hand, in view of (4.11), the Abbe invariant

$$\frac{N_2}{z_2} - \frac{N_2}{R_2} = \frac{N_2'}{z_2'} - \frac{N_2'}{R_2},$$

yields

$$\frac{2}{\gamma} = \frac{f_1}{e - d} - \frac{f_1}{f_1 + d}$$
$$= \frac{f_1}{f} \frac{f}{e - d} - \frac{1}{1 - \beta} = \frac{1}{M} \frac{M}{\alpha + \beta} - \frac{1}{1 - \beta}.$$

Therefore, due to (4.13), we have

$$\gamma = \frac{2(1 - \beta)(\alpha + \beta)}{1 - \alpha - 2\beta} = \frac{2M(1 + \alpha)}{1 - M^2}. \tag{4.14}$$

The Gaussian design is now complete since all of the quantities are expressed in terms of α and M, which are, in turn, obtainable from the initial data d_1, f_1, and f.

4.3 Spherical Aberration and Coma in Cassegrain Telescopes

We have already noted that, if no other optical element is introduced into the scheme, the only way to control the aberrations from the Cassegrain combination is to suppose that S_1 and S_2 are quadric surfaces of revolution. Actually, their conic constants K_1 and K_2 constitute two degrees of freedom that can be used to eliminate or reduce some aberrations.

In order to evaluate the aberration coefficients D_i (see (3.92)–(3.96)), all the quantities that appear in them must be calculated. To avoid having to perform this tedious calculation, the reader can use the notebook **Cassegrain** instead. This notebook proves that the coefficients of spherical aberration and coma take the following forms, respectively:

$$32 f_1^3 D_1 = B_{1,0} - K_1 + B_{1,2} K_2,$$
$$8 f^2 D_2 = B_{2,0} + B_{2,2} K_2, \tag{4.15}$$

4.3. Spherical Aberration and Coma in Cassegrain Telescopes

where

$$B_{1,0} = -\frac{1+M^3+\alpha-M^2(1+\alpha)}{M^3}, \quad B_{1,2} = \frac{(M-1)^3(1+\alpha)}{M^3(1+M)},$$
$$B_{2,0} = \frac{M+M^3+\alpha(1-M^2)}{M^3}, \quad B_{2,2} = \frac{(M-1)^3(M-\alpha)}{M^3(1+M)}.$$
(4.16)

Now, there are many ways to eliminate either spherical aberration or spherical aberration and coma:

- The **Dall–Kirkham** scheme uses a conic primary and a spherical secondary; i.e., $K_2 = 0$. From the conditions $32f^3 D_1 = 0$ and $K_2 = 0$, we get

$$K_1 = B_{1,0}. \tag{4.17}$$

- The **Pressman–Camichel** configuration uses a spherical primary and a conic secondary. Thus, the conditions $8f^3 D_1 = 0$ and $K_1 = 0$ imply

$$K_2 = -\frac{B_{1,0}}{B_{1,2}}. \tag{4.18}$$

- The **classical Cassegrain** combination uses a parabolic primary ($K_1 = -1$) and a hyperbolic secondary. From the conditions $D_1 = 0$ and $K_1 = -1$, we derive

$$K_2 = -\frac{(1+B_{1,0})}{B_{1,2}}. \tag{4.19}$$

- Both coma and spherical aberration vanish in the **Ritchey–Cretien** combination in which both mirrors are conic. From the equations $D_1 = 0$ and $D_2 = 0$, we obtain

$$K_1 = B_{1,0} - \frac{B_{1,2} B_{2,0}}{B_{2,2}}, \quad K_2 = -\frac{B_{2,0}}{B_{2,2}}. \tag{4.20}$$

An analysis of the aberrations obtained with the different combinations listed above shows that coma is greater in Dall–Kirkham or Pressman–Camichel telescopes than in a pure Cassegrain telescope. Moreover, even though it gives excellent performance, the Ritchey–Cretien combination is difficult to realize in practice because it requires two conic mirrors.

The notebook *Cassegrain* allows the reader to evaluate all of the parameters for all of these combinations, as well as the residual third-order aberrations.

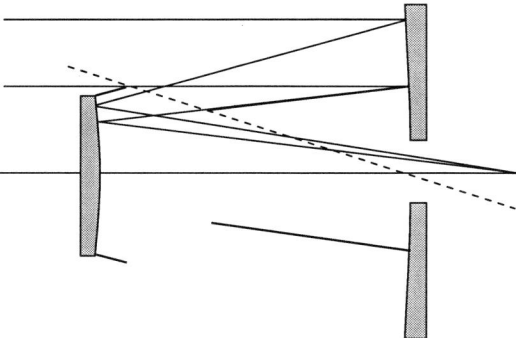

Fig. 4.3 Baffles in a Cassegrain telescope

4.4 Concluding Remarks

Cassegrain telescopes have the following advantages:

- The total focal length of the combination is much longer than the focal length of primary mirror
- The optical system is fully achromatic.

However, they have some disadvantages too:

- Conic surfaces are not easy to make
- The use of an open tube can lead to turbulence phenomena
- Direct light can reach the observer's eye unless an accurate system of baffles is introduced into the tube (see Figure 4.3).

4.5 Examples

In this section we will design a standard Cassegrain telescope and a Ritchey–Cretien telescope using the notebook ***Cassegrain***. Both of these combinations have the following characteristics: the total focal length is 2000 mm, the back distance is 200 mm, and the obstruction factor is 0.30. Moreover, the diameter of the primary conic mirror is 200 mm, and its speed is $F/3$. Finally, we consider a field angle of 0.5°.

4.5. Examples

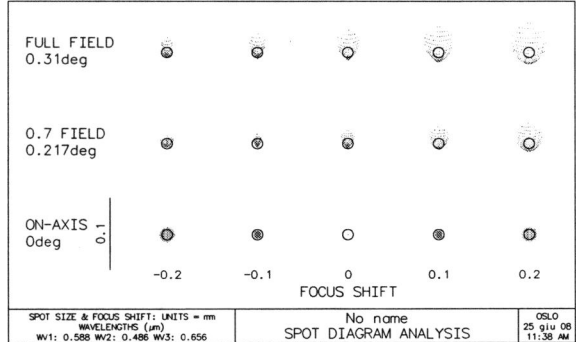

Fig. 4.4 Cassegrain combination

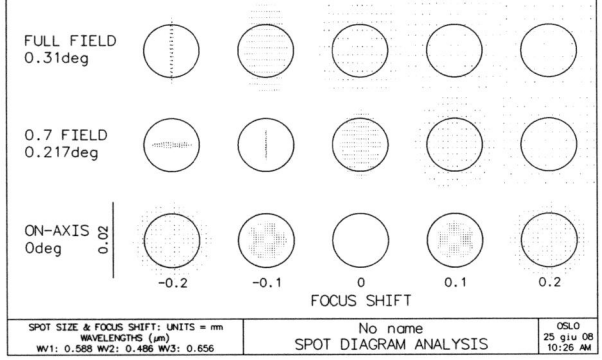

Fig. 4.5 Ritchey–Cretien combination

For these input data, the notebook ***Cassegrain*** gives the following output:

$$\begin{aligned}
\text{Radius of the primary mirror} &= -1200 \\
\text{Radius of the secondary mirror} &= -527.473 \\
\text{Distance between the two mirrors} &= -415.385 \\
\text{Diameter of the secondary mirror} &= 61.54 \\
\text{Focal length} &= 2000 \\
\text{Height of the image} &= 10.82.
\end{aligned}$$

Moreover, for the pure Cassegrain combination, the conic constants K_1 and K_2 of the two mirrors are

$$K_1 = -1, \quad K_2 = -3.449,$$

whereas for the Ritchey–Cretien combination, the values of the two constants become

$$K_1 = -1.08, \quad K_2 = -4.207.$$

The spot diagrams obtained from the software package **OSLO** for the above combinations are shown in Figures 4.4 and 4.5.

In both figures, the circle is the Airy disk and the spot diagram is drawn on a spherical surface with a radius of curvature of -600 mm.

Chapter 5

Cameras for Astronomy

5.1 Introduction

This chapter describes the cameras most commonly used for astronomical photography. The main component in these cameras is a concave mirror S that forms the image of an object at infinity on its focal plane. The image formed by such a system usually exhibits aberrations that are too large to be acceptable. In order to improve this situation, we can:

- Place an aperture stop A a suitable distance s in front of S in order to reduce or eliminate some of the aberrations

- Use a conic mirror S

- Introduce a new optical element at the aperture stop to compensate for the aberrations from S.

All of the cameras considered here are based on the following principle:

The optical element introduced at the aperture stop is almost afocal, and thus compensates for the aberrations from S without modifying its Gaussian behavior. In particular, the image from the whole system is very close to the focal plane of S.

The above considerations indicate that we should study the aberrations of the optical system formed by a conic mirror S with an aperture stop located a distance s from S in great detail. We denote the absolute value of the focal length and the diameter of S by f and D, respectively.

When an afocal element is introduced at the aperture stop, we only need to add the aberration coefficients of this element to the corresponding coefficients of S.

It is evident that all of the results derived here are only acceptable in the third-order approximation. In particular, the **aperture** or the

5.2 Aberrations for a Single Mirror

We consider an optical system comprising a single mirror S with a radius of curvature $R = -2f < 0$ and a conic constant K. Moreover, we suppose that an aperture stop A is placed in front of S at a distance $s = -\delta f$. In the preceding section it was stated that all of the cameras considered in this chapter are obtained by inserting a particular afocal optical element into the plane of the aperture stop. To avoid modifying the signs of the quantities that appear in the formulae, it is initially supposed that the aperture stop is the rim of a plate with planar and parallel faces, where the thickness of the plate can be neglected. Clearly, this optical element has no effect on the aberrations of the primary mirror. All the formulae derived here can be obtained using the notebook *SCamera*.

Since the object is at infinity and $N = -N' = 1$, we can derive the distance of the image from the mirror from Abbe's invariant (2.8):

$$z' = -f. \tag{5.1}$$

Using (2.7) and (3.59), we also obtain

$$w' = \frac{f\delta}{1-\delta}, \tag{5.2}$$

$$M_e = \frac{1}{1-\delta}, \quad c = 1, \quad d = 0. \tag{5.3}$$

Moreover, the coefficients (3.42)–(3.46) become

$$C_1 = -\frac{K+1}{32f^3}, \tag{5.4}$$

$$C_2 = \frac{1}{4f^2}, \tag{5.5}$$

$$C_3 = 0, \tag{5.6}$$

$$C_4 = -\frac{1}{2f}, \tag{5.7}$$

whereas the relations (3.61)–(3.65) assume the form

$$C_1^* = -\frac{K+1}{32f^3}, \tag{5.8}$$

5.3. Schmidt Cameras

$$C_2^* = \frac{1}{8f^2}\left(2 - (K+1)\delta\right), \tag{5.9}$$

$$C_3^* = \frac{\delta}{16f}\left(4 - (K+1)\delta\right), \tag{5.10}$$

$$C_4^* = -\frac{1}{16f}\left(8 - 12\delta + 3(K+1)\delta^2\right), \tag{5.11}$$

$$C_5^* = -\frac{1}{8f}\delta\left(8 - 6\delta + (K+1)\delta^2\right). \tag{5.12}$$

Since we have just one surface, the coefficients of total spherical aberration and coma, $D_{S,1}$ and $D_{S,2}$, coincide with (5.8) and (5.9):

$$32f^3 D_{S,1} = -(K+1), \tag{5.13}$$

$$8f^2 D_{S,2} = 2 - (K+1)\delta, \tag{5.14}$$

whereas the total coefficients of astigmatism and field curvature are:

$$8f(D_{S,4} - D_{S,3}) = 8f(C_4^* - C_3^*) = -\left(4 - 4\delta + (K+1)\delta^2\right), \tag{5.15}$$

$$D_{S,4} - 3D_{S,3} = C_4^* - 3C_3^* = -\frac{1}{2f}. \tag{5.16}$$

Moreover, the Petzval, sagittal, and meridional radii are, respectively, given by the relations

$$R_p = -f, \tag{5.17}$$

$$R_s = -\frac{4f}{\delta(-4 + \delta(K+1))}, \tag{5.18}$$

$$R_t = \frac{4f}{8 - 12\delta + 3(K+1)\delta^2}. \tag{5.19}$$

5.3 Schmidt Cameras

Equations 5.8–5.12 show that, if the stop is located at the mirror S ($\delta = 0$) and S is a sphere ($K = 0$), all of the third-order aberrations are present except for the distortion. In contrast, if the aperture stop is placed at the center of curvature of S ($\delta = 2$) and $K = 0$, then we get:

$$D_{S,1} = -\frac{1}{32f^3}, \tag{5.20}$$

$$D_{S,2} = 0, \tag{5.21}$$

$$D_{S,4} = D_{S,3}, \tag{5.22}$$

$$D_{S,5} = 0. \tag{5.23}$$

Focal length	Stop diameter
400 mm	74.26 (5.4)
500 mm	86.17 (5.8)
600 mm	97.31 (6.1)
700 mm	107.8 (6.5)
800 mm	117.8 (6.8)
900 mm	127.5 (7.1)
1000 mm	136.8 (7.3)
1200 mm	154.5 (7.7)
1500 mm	179.2 (8.4)

Table 5.1 Focal length, diameter, and speed of a spherical mirror satisfying the photographic criterion

Using $D_{S,4} = D_{S,3}$ we find that the sagittal and tangential curvatures are equal and coincide with Petzval's curvature $2/R$. This conclusion can also be derived by inspecting (5.17)–(5.19).

In other words, when the aperture stop is located at the center of curvature, coma, astigmatism and distortion vanish, whereas the spherical aberration and field curvature survive. All of these results suggest that we should use a single spherical mirror with the aperture stop located at its center of curvature, provided that its speed is such that the aberration spot has a size that allows us to use it. More precisely, it is possible to prove that the diameter D of S must satisfy the condition

$$D = 2(4R^2 d_b)^{1/3}, \qquad (5.24)$$

where d_b is the diameter of the allowed spot. For photographic applications, $d_b = 0.025$ mm, and the data shown in Table 5.1 are obtained.

In this table, the numbers in parentheses in the second column denote the ratio of the diameter to the focal length $f = |R|/2$ of S (i.e., the speed of S). It is evident that the length $2f$ of the optical combination is only convenient for small diameters of S.

Alternatively, the aperture stop could be positioned at the center of curvature of S along with a suitable refractive plate P that has the task of compensating for the spherical aberration of S. In this case, the aperture stop will be the rim of P.

The excellent configuration obtained in this case (see Figure 5.1) was invented by the German optician Bernard Schmidt. At the beginning of the 1920s he created a very fast optical system that was able to eliminate almost all aberrations. Because of its characteristics and its remarkable optical performance, this configuration is employed when photographing wide star fields.

5.3. Schmidt Cameras

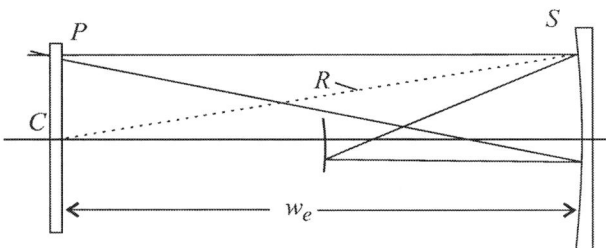

Fig. 5.1 Schmidt camera

This configuration essentially consists of two components: a concave spherical mirror S and a *specially shaped* corrector plate P placed at the center of curvature C of S. This optical design uses the symmetry of the system to eliminate almost all off-axis aberrations, *even those of higher order*, and it entrusts the corrector with the task of eliminating the spherical aberration from the primary mirror. The image is formed on a spherical surface of radius $-f$, onto which a film is placed.

The corrector (the rim of which is the aperture stop) has a much greater focal length than that of S; in other words, the system is almost afocal. This implies that Gaussian calculations do not need to consider its presence. More precisely, we assume that:

- The first surface of P is planar
- Its thickness is very small compared with the radius of curvature of the mirror S
- The profile of the second face is given by (1.28), where the radius of curvature R_c at the vertex is much greater than R.

The above hypotheses allow us to conclude that

$$z_1 = z_1' = z_2 = z_2' = \infty,$$
$$w_1 = w_2 = 0,$$
$$M_{e,1} = M_{e,2} = 1.$$

$$C_{1,1}^* = 0, \quad C_{i,2}^* = C_{i,3}^* = C_{i,4}^* = C_{i,5}^* = 0, \quad i = 1, 2, \tag{5.25}$$
$$C_{2,1}^* = D_{P,1} = a_4(N-1), \tag{5.26}$$

where N is the refractive index of the corrector (see (3.34)). Consequently, all of the coefficients D_i for the whole combination coincide with the coefficients $D_{S,i}$ for the mirror (see (5.9)), except for D_1, which is given by

$$D_1 = D_{P,1} + D_{S,1} = a_4(N-1) - \frac{1}{32f^3}. \tag{5.27}$$

The value of a_4 that allows us to eliminate the spherical aberration is then

$$a_4 = \frac{1}{32(N-1)f^3}. \tag{5.28}$$

Because of our hypotheses, the equation for a section of P is (see (1.29))

$$z = \frac{1}{2R_c}y^2 + a_4 y^4, \tag{5.29}$$

where

$$a_4 = \frac{1}{32(N-1)f^3}. \tag{5.30}$$

Up to now, the radius of curvature R_c of the corrector P at the vertex has not been determined. We can do this by imposing that $z'(\alpha r) = 0$, where r is the radius of P and α is a coefficient that lies within the interval $[0, 1]$. In this way, we can easily derive

$$R_c = -\frac{8(N-1)f^3}{\alpha^2 r^2}. \tag{5.31}$$

Finally, the radius R_c and the coefficient a_4 in (5.14) are given by

$$R_c = -\frac{8(N-1)f^3}{\alpha^2 r^2}, \quad a_4 = \frac{1}{32(N-1)f^3}. \tag{5.32}$$

Figure 5.2 shows different profiles of P upon varying the value of α. The first corresponds to the value $\alpha = 0$, which, in turn, implies $R_c = \infty$ or $a_2 = 0$. In this case, the focal length of the mirror S is not modified by the corrector. The second corresponds to a corrector that has the same thickness at both the edge and the center. The value $\alpha = 0.866$ is preferred since it gives the best chromatic correction.

Note that all of the above formulae are only valid given the approximations listed at the beginning of this chapter, and they do not take into account the small contributions of the curvature of the second surface of the corrector and its thickness to the aberrations. They do however yield good results provided that the main mirror has a speed of less than $f/3$.

After focusing mainly on the advantages of this camera in this section, we should also list its disadvantages. The first of them is its length, which is equal to the radius of curvature of the primary mirror. Another is the difficulty involved in generating the correct profile for the corrector, especially if a sixth-order profile is requested in order to control higher-order spherical aberrations when a high-speed principal mirror is employed. Other solutions that overcome these problems—although not without paying a price—are discussed in the following sections of this chapter.

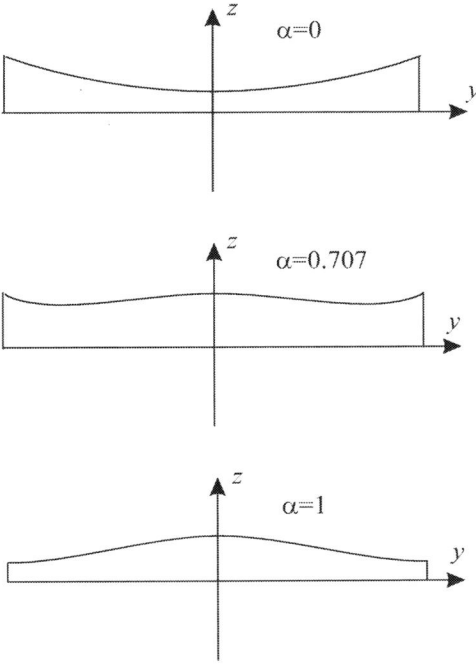

Fig. 5.2 Different profiles for the Schmidt corrector

5.4 Examples

Suppose that $\alpha = 0.8666$, the mirror S has a radius of curvature $R = -1600$ mm, the corrector P has a diameter of 200 mm (speed $f/4$) and is made from BK7 glass, which has a refractive index for green light of $N = 1.518722$. Using the above formulae or the notebook *SCamera*, we can derive:

$$a_2 = -283296, \quad a_4 = 1.1766 \times 10^{-10}.$$

Similarly, the corrector associated with a mirror that has a radius of curvature $R = -1200$ (speed $f/3$) has the following characteristics:

$$R_c = -119521, \quad a_4 = 2.789 \times 10^{-10}.$$

When a camera with a mirror that has a speed of $f/2.5$ is considered, we obtain

$$R_c = -69164, \quad a_4 = 4.82 \times 10^{-10}.$$

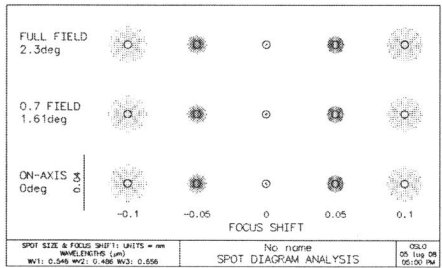

Fig. 5.3 Spot diagram for a Schmidt camera with a speed of f/4

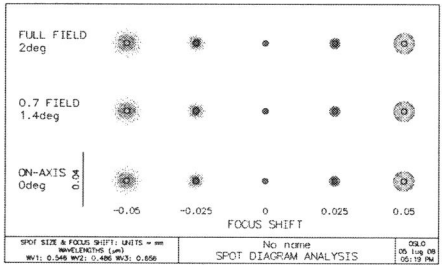

Fig. 5.4 Spot diagram for a Schmidt camera with a speed of f/3

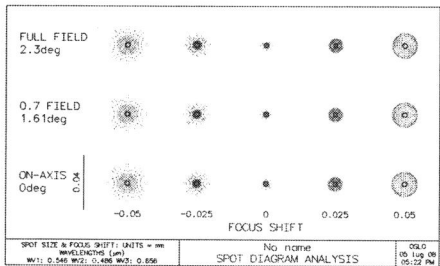

Fig. 5.5 Spot diagram for a Schmidt camera with a speed of f/2.5

5.5. Wright Cameras

We can obtain spot diagrams that can be used to evaluate these cameras using professional software packages like **OSLO** or **ATMOS**. Figures 5.3–5.5 were obtained using **OSLO**. We can see that excellent correction is achieved on a spherical image surface with a radius of curvature of $-f$ when the speed is $f/4$ or $f/3$. When the speed is $f/2.5$, the performance of the camera is still acceptable, but it is clear that we must resort to more accurate methods if we wish to make faster cameras. More precisely, we must

- Take into account the aberrations due to the finite radius and thickness of the second surface of the corrector

- Balance the third-order and fifth-order spherical aberrations; this becomes important for mirrors that are faster than $f/2.5$.

5.5 Wright Cameras

In this section we show that we can have a shorter camera provided that we are willing to accept reduced image quality.

We start by noting that any displacement of the corrector from the center of curvature of the primary mirror introduces coma and astigmatism. A **Wright camera** contains the same components as a Schimdt camera but the corrector is closer to the mirror, which is now assumed to be *conic*. Consequently, there are only two free parameters for this combination, the coefficient a_4 of the corrector and the conic constant K of the mirror. Therefore, we can hope to control the spherical aberration and coma by choosing these parameters appropriately. A Wright camera is an aplanatic combination; in other words, it is free from spherical aberration and coma.

Let us consider a concave conic mirror S and a Schmidt corrector P that is situated a distance $\delta f < |R|$ in front of S. If a_4 denotes the aspherical constant of P, N its refractive index, f the focal length of S, and K the conic constant of S, it is straightforward to verify that the spherical aberration, sagittal coma and astigmatism coefficients of this combination (see (5.6) and the notebook **WCamera**) are:

$$32f^3 D_1 = -(K+1) + 32f^3 a_4(N-1), \tag{5.33}$$
$$8f^2 D_2 = 2 - (K+1)\delta, \tag{5.34}$$
$$8f(D_4 - D_3) = -\left(4 - 4\delta + (K+1)\delta^2\right). \tag{5.35}$$

The expressions (5.33)–(5.35) show that, *for any chosen distance δf between the corrector and the primary mirror, the values of a_4 and K that* *allow spherical aberration and coma to be eliminated satisfy the following system:*

$$32a_4(N-1)f^3 - K - 1 = 0, \qquad (5.36)$$
$$2 - (K+1)\delta = 0. \qquad (5.37)$$

The solution to this system is

$$a_4 = \frac{1}{16f^3(N-1)\delta}, \quad K = -1 + \frac{2}{\delta}. \qquad (5.38)$$

These relations prove that the conic constant K of S and the fourth-order coefficient of the corrector P depend only on the stop position. Finally, the astigmatism coefficient corresponding to these values reduces to

$$D_4 - D_3 = \frac{\delta - 2}{4f}. \qquad (5.39)$$

We note that the astigmatism and the conic constant K tend to zero when δ tends to 2 (the corrector is situated at the center of curvature of S). However, for $\delta = 2$, the Wright camera becomes a Schmidt camera. On the other hand, these quantities increase when the corrector is closer to S. In particular, K assumes very high values when $\delta < 1$; consequently, the mirror becomes more and more difficult to produce. We can therefore conclude that the most convenient values of δ belong to the interval 1.2–1.7. These values allow the tube to be shortened with respect to Schmidt's combination by 40–25%, respectively.

Moreover, due to the presence of astigmatism, the surface of the best focus does not coincide with the sphere with the Petzval radius, but rather with a sphere that has a radius that lies between the sagittal and meridional radii.

Some examples of Wright cameras can be found in the notebook **WCamera**. These examples show that this configuration is more difficult to optimize than Schmidt's one. In fact, when the speed increases, a higher-order astigmatism appears, which reduces the efficiency of the third-order correction.

5.6 Houghton Cameras

Another way to make an aplanatic camera using *spherical surfaces* only is given by **Houghton's combination** (see Figure 5.6). This is realized by replacing the aspherical Schmidt corrector with a pair H of spherical

5.6. Houghton Cameras

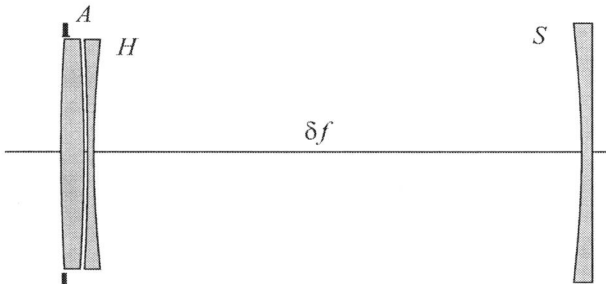

Fig. 5.6 Houghton camera

lenses, made from the same glass, whose radii of curvature R_i, $i = 1, \ldots, 4$, satisfy the conditions

$$R_1 = -R_3, \quad R_2 = -R_4. \tag{5.40}$$

It is straightforward to show that, *for any wavelength*, an afocal system is obtained *when the thickness of these lenses goes to zero*. In fact, the focal length of two adjacent thin lenses is (see **HCamera**):

$$f = \frac{1}{(c_1 - c_2 + c_3 - c_4)(N - 1)} \tag{5.41}$$

(where $c_i = 1/R_i$, $i = 1, \ldots, 4$, denote the curvatures of the corrector surfaces), which goes to infinity when (5.40) holds.

We therefore reach the following conclusions:

- H does not modify the Gaussian description of the primary mirror S, so the aberration coefficients for the whole combination are obtained by adding the coefficients from (5.6) to the corresponding coefficients for the two lenses

- No axial chromatic aberration is introduced by the corrector provided that *we can neglect the thickness of the two lenses*.

The expressions for the spherical aberration, coma, and astigmatism coefficients of S are obtained via (5.33)–(5.35) for $a_4 = K = 0$:

$$D_{S,1} = -\frac{1}{32f^3}, \tag{5.42}$$

$$D_{S,2} = \frac{1}{8f^2}(2 - \delta), \tag{5.43}$$

$$D_{S,4} - D_{S,3} = -\frac{1}{8f}(2 - \delta)^2. \tag{5.44}$$

If we assume that the thickness of the lens comprising Houghton's corrector is zero, application of the notebook **HCamera** yields the following aberration coefficients[1]

$$D_{H,1} = (c_1 - c_2)^2(c_1 + c_2)\frac{(N+1)(N-1)^2}{4N}, \qquad (5.45)$$

$$D_{H,2} = -(c_1^2 - c_2^2)\frac{(N^2-1)}{2N}, \qquad (5.46)$$

$$D_{H,4} - D_{H,3} = 0, \qquad (5.47)$$

where N is the refractive index of the lenses and $c_1 = 1/R_1$ and $c_2 = 1/R_2$ are their curvatures.

The system is aplanatic when the curvatures verify the following equations (and $D_{S,1}$ and $D_{S,2}$ have been assigned):

$$D_{H,1} + D_{S,1} = 0, \quad D_{H,2} + D_{S,2} = 0. \qquad (5.48)$$

When the expressions (5.45)–(5.47) are inserted into (5.48), the following sixth-degree system is obtained:

$$(c_1 - c_2)^2(c_1 + c_2)\frac{(N+1)(N-1)^2}{4N} = -D_{S,1}, \qquad (5.49)$$

$$(c_1^2 - c_2^2)\frac{(N^2-1)}{2N} = D_{S,2}. \qquad (5.50)$$

Solving this yields the curvatures of the two lenses needed to eliminate spherical aberration and coma. It is very important to note that this system is equivalent to a linear one, so these derived solutions (for the curvatures) are unique. In fact, from (5.50), we have

$$c_1^2 - c_2^2 = \frac{2ND_{S,2}}{N^2-1}, \qquad (5.51)$$

and substituting this expression into $(5.49)_1$ gives

$$c_1 - c_2 = -\frac{2D_{S,1}}{(N-1)D_{S,2}}.$$

Using this formula, the following linear system can be derived:

$$c_1 - c_2 = -\frac{2D_{S,1}}{(N-1)D_{S,2}}, \qquad (5.52)$$

$$c_1 + c_2 = -\frac{ND_{S,2}}{(N-1)D_{S,1}}, \qquad (5.53)$$

---

[1] The analysis of a Houghton camera presented in this section can be found in [26].

5.6. Houghton Cameras

which has the two unique solutions

$$c_1 = -\frac{2(N+1)D_{S,1} + N(N-1)D_{S,2}^3}{2(N^2-1)D_{S,1}D_{S,2}}, \quad (5.54)$$

$$c_2 = \frac{2(N+1)D_{S,1}^2 - N(N-1)D_{S,2}^3}{2(N^2-1)D_{S,1}D_{S,2}}. \quad (5.55)$$

Finally, using (5.6) to express $D_{S,1}$ and $D_{S,2}$, and recalling the relations $R_i = 1/c_i$, we obtain the radii of the doublet:

$$\begin{aligned} R_1 &= f\frac{4(N^2-1)(2-\delta)}{(N+1) + N(N-1)(2-\delta)^3}, \\ R_2 &= -f\frac{4(N^2-1)(2-\delta)}{(N+1) - N(N-1)(2-\delta)^3}. \end{aligned} \quad (5.56)$$

Moreover, the astigmatism is absent when (see (5.7)):

$$\begin{aligned} (D_{H,4} + D_{S,4}) - (D_{H,3} + D_{S,3}) &= \\ D_{S,4} - D_{S,3} &= -\frac{1}{8f}(2-\delta)^2 = 0, \end{aligned} \quad (5.57)$$

in other words, when the aperture stop A is placed at the curvature center of S ($\delta = 2$). It increases when A is closer to the mirror S. Therefore, the astigmatism of the whole combination coincides with the astigmatism of the primary mirror, since the doublet does not contribute to it, at least when the thickness of the lenses of the doublet is negligible.

We could consider eliminating the astigmatism by positioning the aperture stop at the center of curvature for the mirror. However, when $\delta \to 2$, the radii R_1 and R_2 go to zero because of (5.56). Consequently, higher-order aberrations appear and it becomes impossible to produce lenses of the required diameter. When $\delta > 2$, the radii R_1 and R_2 again increase, changing their signs, but the tube is longer. Figure 5.7 shows all of these conclusions graphically and suggests that we should choose $\delta \in [0.7, 1.2]$.

When $\delta \simeq 1.2$, the radii R_1 and R_2 are still sufficiently high and the astigmatism is acceptable for photographic purposes. Moreover, the tube is shorter than in an equivalent Schmidt camera.

We now consider the case of a corrector with different radii where the radius of the fourth surface is such that the corrector is afocal (see 5.41). In the notebook **HCamera**, it is shown that the relations (5.49) and (5.50)

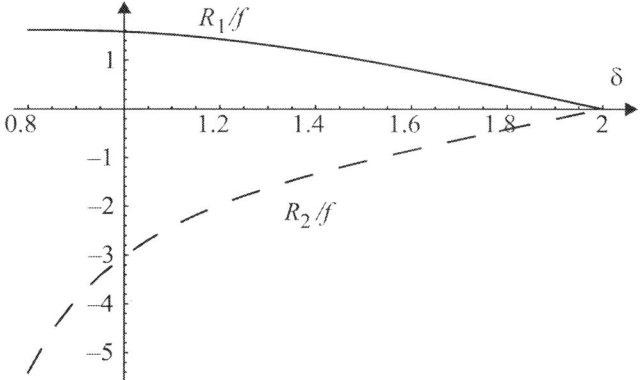

Fig. 5.7 Radii R_1/f and R_2/f versus δ

can be replaced with the following:

$$(c_1 - c_2)(c_2 - c_3)(-\frac{1}{8}(3c_1 - 2c_2 + c_3)(N+1)-$$
$$\frac{1}{4N}(2c_1 - c_2 + c_3) - \frac{1}{4}(c_1 - c_2)N^2) = -D_{S,1}, \quad (5.58)$$
$$-(c_1 - c_2)(c_2 - c_3)\frac{N^2-1}{2N} = -D_{S,2}.$$

Also, in this case, the doublet does not contribute to the astigmatism, since
$$D_{H,4} - D_{H,3} = 0.$$

Again, this system looks to be a sixth-degree system. However, it is possible to verify that it is equivalent to a second-degree system. First, we adopt the notation
$$x = \frac{c_1}{c_3} = \frac{R_3}{R_1}, \quad y = \frac{c_2}{c_3} = \frac{R_3}{R_2},$$
so that the system becomes

$$(y-1)(x-y)\left(-\frac{1}{8}(3x - 2y + 1)(N+1)+\right.$$
$$\left.\frac{2x - y + 1}{4N} + \frac{N^2}{4}(x-y)\right) = -D_{S,1}R_3^3, \quad (5.59)$$
$$(x-y)(y-1)\frac{N^2-1}{2N} = D_{S,2}R_3^2.$$

5.6. Houghton Cameras

Substituting the relation

$$(x-y)(y-1) = \frac{2N}{N^2-1} D'_{S,2} R_3^2 \tag{5.60}$$

deduced from the second equation of (5.59) into (5.59)$_1$, we obtain the second-degree system

$$-\frac{1}{8}(3x - 2y + 1) + \frac{2x - y + 1}{4N} +$$
$$\frac{N^2}{4}(x - y) = -\frac{N^2-1}{2N} \frac{D_{S,1}}{D_{S,2}} R_3, \tag{5.61}$$
$$(x-y)(y-1) = \frac{2N}{N^2-1} D_{S,2} R_3^2.$$

Taking (5.6) into account, we finally get

$$a_1 x + a_2 y = d_1,$$
$$(x-y)(y-1) = d_2, \tag{5.62}$$

where

$$r_3 = \frac{R_3}{f}, \quad a_1 = \frac{4 - 3N - 3N^2 + 2N^3}{8N},$$
$$a_2 = \frac{-1 + N + N^2 - N^3}{4N}, \tag{5.63}$$
$$d_1 = -\frac{1}{8} - \frac{1}{4N} + \frac{N}{8} + \frac{(N^2-1)r_3}{8N(2-\delta)},$$
$$d_2 = \frac{N}{4(N^2-1)} (2-\delta) r_3^2.$$

System (5.62) has the following two solutions:

$$r_1 = 2r_3 \frac{a_1(a_1 + a_2)}{-a_1(a_2 - 2d_1) + a_2(-a_2 + d_1 + \sqrt{\Delta})},$$
$$r_2 = 2r_3 \frac{a_1 + a_2}{a_1 + a_2 + d_1 - \sqrt{\Delta}}, \tag{5.64}$$

or

$$r_1 = -2r_3 \frac{a_1(a_1 + a_2)}{a_1(a_2 - 2d_1) + a_2(a_2 - d_1 + \sqrt{\Delta})},$$
$$r_2 = 2r_3 \frac{a_1 + a_2}{a_1 + a_2 + d_1 + \sqrt{\Delta}}. \tag{5.65}$$

These are real provided that the discriminant

$$\Delta = (a_1 + a_2 + d_1)^2 - 4(a_1 + a_2)(d_1 + a_1 d_2) > 0. \tag{5.66}$$

i.e., that the discriminant is positive.

116   Chapter 5. Cameras for Astronomy

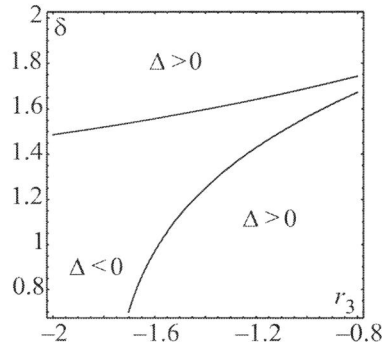

Fig. 5.8 Sign of Δ in the plane (r_2, δ)

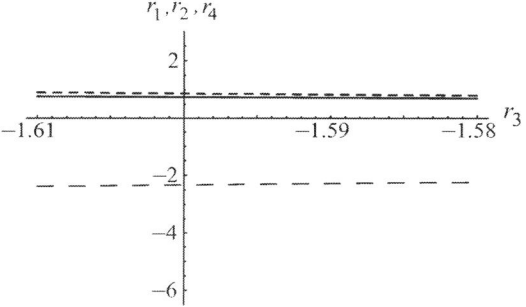

Fig. 5.9 $\delta = 0.8$, solution (5.64)

Substituting (5.63) into (5.66) and (5.65), we can determine all of the acceptable values of r_3 for a given δ. Figure 5.8 shows the subset of the plane (r_3, δ) where (5.66) is valid.

Using **Mathematica**, we can analyze the behavior of the radii r_1 and r_2 upon varying r_3 in the set I_δ of acceptable values for a given δ. Figures 5.9–5.10 plot the variables r_1, r_2 and r_4 versus r_3 for $\delta = 0.8$. More precisely, the first figure refers to solution (5.64) and the second to (5.65). The interval chosen for δ starts from the lower bound of I_δ (see **HCamera**).

In Figures 5.9–5.10, the continuous line refers to r_1, the long-dashed line to r_2 and the short-dashed line to r_4.

Finally, Figures 5.11–5.12 have exactly the same meaning as Figures 5.9–5.10 but correspond to $\delta = 1.1$ instead.

An analysis of the plots leads us to the following conclusions:

- The highest and thus the most convenient values of r_1, r_2, r_3 and r_4 are obtained for $\delta \in [0.7, 1]$.

5.6. Houghton Cameras

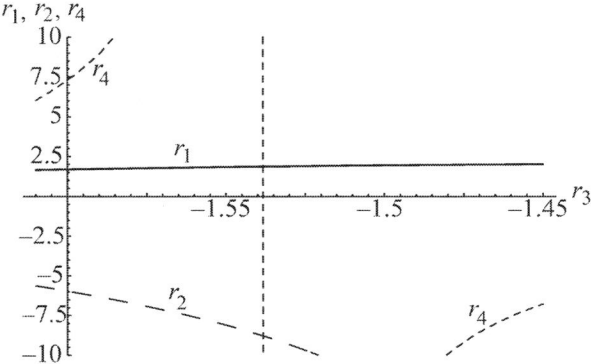

Fig. 5.10 $\delta = 0.8$, solution (5.65)

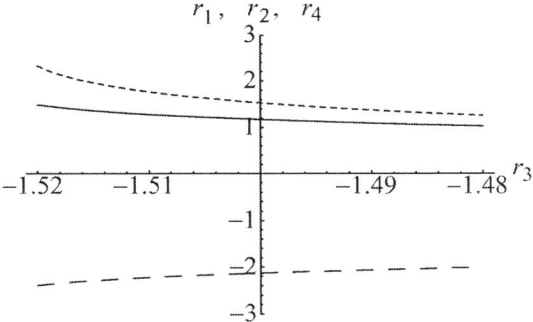

Fig. 5.11 $\delta = 1.1$, solution (5.64)

- When the assigned δ falls within this interval, solution (5.64) and both r_3 and r_4 are greatest at the lower bound of I_δ, whereas solution (5.65) and both r_3 and r_4 must be evaluated at 0.8–0.9 times the lower bound. For instance, if $\delta = 0.8$, from Figure 5.8 we can derive $I_\delta = [-1.65, -0.8]$, so solution (5.64) must be evaluated at $r_3 = -1.65$, whereas solution (5.65) should be evaluated at $r_3 = -1.5$ (see Figures 5.9 and 5.10, which refer to solutions (5.64) and (5.65), respectively).

- In any case, the second solution leads to larger radii.

The program **HCamera2** in the notebook **HCamera** allows us to design Houghton cameras with different radii. Good results are obtained for mirrors that are slower than $f/3$.

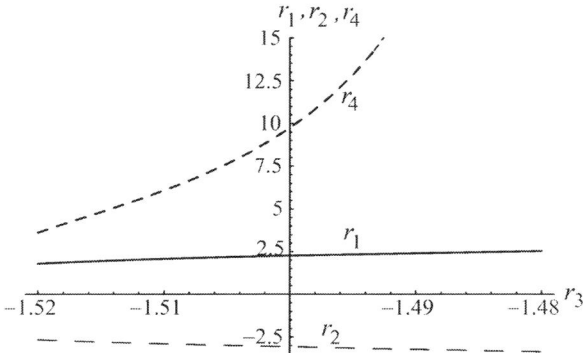

Fig. 5.12 $\delta = 1.1$, solution (5.65)

If we are willing to introduce aspherical surfaces into the system, it then becomes possible to eliminate the astigmatism by adopting a conic mirror S. We have already noted that the corrector does not introduce astigmatism (see notebook **HCamera**) when it is afocal and very thin. Consequently, the total astigmatism of the whole combination corresponds to the astigmatism of the primary mirror. Taking into account the first equation of (5.15), we can then conclude that *it is possible to eliminate this aberration for any aperture stop position; that is, for any $\delta > 0$*, provided that the conic constant of S is

$$K = -\frac{(\delta-2)^2}{\delta^2}. \tag{5.67}$$

When the conic constant of S is this value, the spherical aberration and coma coefficients (5.6) become

$$D_{S,1} = \frac{1-\delta}{8f^3\delta^2}, \quad D_{S,2} = \frac{\delta-2}{4f^2\delta}. \tag{5.68}$$

Finally, if these expressions are inserted into (5.56), the following formulae for the radii of a symmetric doublet can be derived:

$$R_1 = \frac{4\delta(N^2-1)(\delta-2)(\delta-1)}{-2(\delta-1)^2 + N^2\delta(\delta-2)^3 - N(2-12\delta+14\delta^2-6\delta^3+\delta^4)},$$

$$R_2 = \frac{4\delta(N^2-1)(\delta-2)(\delta-1)}{2(\delta-1)^2 + N^2\delta(\delta-2)^3 - N(2+4\delta-10\delta^2+6\delta^3-\delta^4)}. \tag{5.69}$$

5.6. Houghton Cameras

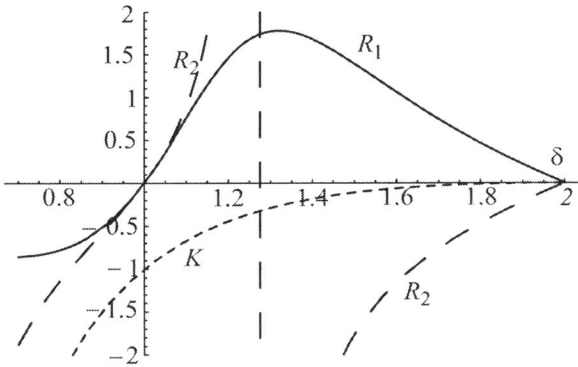

Fig. 5.13 Plots of R_1, R_2, and K versus δ

It is important to note that, when $\delta = 1$ (i.e., the aperture stop is positioned on the focal plane of S), the spherical coefficient of S is equal to zero. Consequently, in order to eliminate the spherical aberration of the whole combination, we must have $D'_{H,1} = 0$, which in turn implies $c_1 = c_2$ (see (5.45)). On the other hand, for these values of c_1 and c_2, the coma coefficient from (5.46) of the corrector vanishes and it is impossible to eliminate coma from the combination. Moreover, the mirror S becomes more and more conic when $\delta \to 0$ (see Figure 5.13).

We can now state the following:

- When $\delta \in [1.4, 2]$, the radii of the corrector surfaces become smaller and so higher-order aberrations appear. Convenient values of δ lie in the interval $[1.2, 1.4]$.

The program **HCamera3** in **HCamera** uses all of the above formulae to derive the radii of curvature of Houghton's corrector and the conic constant of the mirror.

All of the results discussed in this section can summarized as follows:

- If a conic mirror S is used, we can design Houghton cameras with speeds of up to $f/2.5$ that are about 40% shorter than the corresponding Schmidt cameras.

- Using only spherical surfaces, we can design good cameras with speeds of up to $f/3.5$ if the lenses of the corrector have paired radii, or up to $f/3$ if the lenses of the corrector have different radii. Moreover, these cameras are about 50% shorter than the equivalent Schmidt cameras.

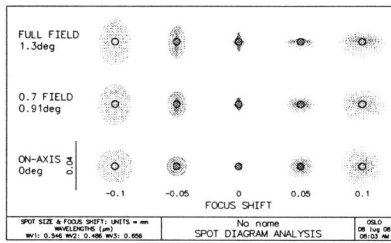

Fig. 5.14 Houghton camera with a speed of $f/4$

5.7 Examples

In this section we discuss some Houghton cameras realized with the notebook **HCamera**, and their spot diagrams. In all of the examples the corrector has a diameter of 200 mm.

We start with a Houghton camera of the first type, with a primary mirror of speed $f/4$. The notebook **HCamera1** then gives

$$R_1 = 1288, \quad R_2 = -5332.4,$$

whereas the distance between the corrector and primary mirror is 600 mm. The spot diagram on a sphere of -1000 mm and for a view angle of $1.3°$ is given in Figure 5.14.

Figure 5.15 shows the spot diagram for a Houghton camera of the second type, with a speed of $f/3$. The data for this camera, obtained using the program **HCamera2**, are:

$$R_1 = 2638.5, \quad R_2 = -1608.8, \quad R_3 = -815.23, \quad R_4 = -4423,$$

with a distance between the corrector and primary mirror of 750 mm and a view angle of $1.7°$.

The spot diagram was obtained on a sphere with a radius of -800 mm.

Finally, Figure 5.16 shows the spot diagram for a Houghton camera with a conic primary mirror of speed $f/2.7$ on a spherical surface of radius -540. The program **HCamera3** gives

$$R_1 = 955.7, \quad R_2 = -8830,$$

with a distance between the corrector and primary mirror of 700 mm and a view angle of $2°$.

5.8. Maksutov Cameras

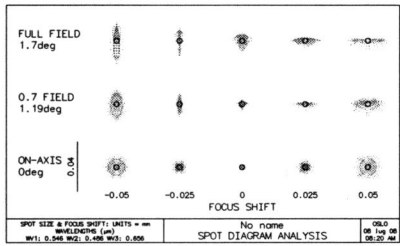

Fig. 5.15 Houghton camera with a speed of $f/3$

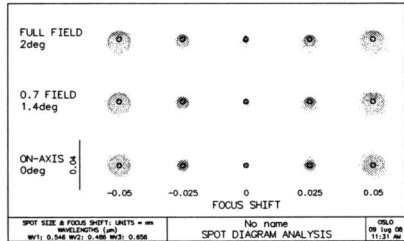

Fig. 5.16 Houghton camera with a speed of $f/2.7$

5.8 Maksutov Cameras

In a Maksutov camera, a meniscus C_M is placed in front of a concave spherical mirror S, so all the surfaces of interest are spherical (see Figure 5.17).

The aperture stop A coincides with the rim of C_M. In his original paper, Maksutov proposed the following empirical formulae for a photographic camera when the speed $\mathbb{A}$ of S and the diameter D of C_M are assigned and C_M is made from BK7 glass:

$$r_1 = -0.612\mathbb{A}^{0.66}D, \quad r_2 = (-0.612\mathbb{A}^{0.66} - 0.0565 - 0.007\mathbb{A})D,$$
$$\Delta = 0.1D, \quad R = -2.107\mathbb{A}^{0.983}D, \quad h = 1.11\mathbb{A}^{1.14}D, \tag{5.70}$$

where r_1, r_2, and Δ denote the radii of curvature for the two surfaces of the corrector and its thickness, respectively. Moreover, R and h are the radius of curvature of S and its distance from C_M, respectively. These formulae allow us to determine the radii of the meniscus C_M and the primary mirror S when the speed $\mathbb{A}$ and diameter D of S are given.

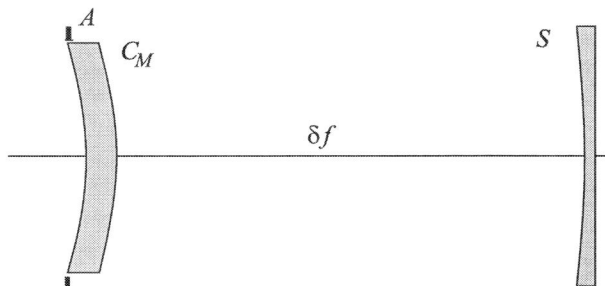

Fig. 5.17 Maksutov camera

However, a camera based on these formulae exhibits consistent third-order spherical aberration. Moreover, due to the strong curvature of the faces of the meniscus, there is also residual fifth-order spherical aberration.

The notebook ***MCamera*** is based on the following considerations.[2] First, the corrector C_M has a very long focal length compared with the focal length of the primary mirror S. This means (see Section 5.1) that the aberration coefficients of the whole system are obtained, to a good approximation, by adding the aberration coefficients for the meniscus to those for S and noting that the aperture stop is located at a distance δf in front of S.

According to (5.36) or, equivalently, ***MCamera***, the spherical aberration coefficient of the spherical mirror S is given by

$$D_{S,1} = -\frac{1}{32f^3}. \tag{5.71}$$

We use the following notation for the meniscus C_M: N is its refractive index, Δ is its thickness, $s = r_1 - r_2$, and D is its diameter. The spherical aberration coefficient of corrector is then (see ***MCamera***):

$$D_{M,1} = \frac{N-1}{8N^3 r_1^4 (r_1-s)^3}[(r_1+(N-1)(s-\Delta))^2(N(r_1-\Delta)+\Delta) \\ (r_1+(N-1)(s+Ns-N\Delta)-r_1 N(r_1-s)^3]. \tag{5.72}$$

Finally, the condition that must hold in order to achieve an axially achromatic meniscus is:

$$\frac{dz_2'}{dN} = 0. \tag{5.73}$$

---

[2]The analysis provided in this section can be found in [27].

5.8. Maksutov Cameras

Since (see **MCamera**)

$$z'_2 = \frac{-Nr_1^2 + Nsr_1 - \Delta r_1 + N\Delta r_1 + s\Delta - Ns\Delta}{(N-1)(Ns + \Delta - N\Delta)}, \qquad (5.74)$$

(5.73) becomes:

$$\begin{aligned}&(N^2 s + \Delta - N^2\Delta)r_1^2 - \\ &(N^2 s^2 + 2s\Delta - 2Ns\Delta - \Delta^2 + 2N\Delta^2 - N^2\Delta^2)r_1 + \\ &(s^2 - 2Ns^2\Delta + N^2 s^2\Delta - s\Delta^2 + 2Ns\Delta^2 - N^2 s\Delta^2) = 0.\end{aligned} \qquad (5.75)$$

The roots of this equation are:

$$s = r_1, \quad s = \frac{(N^2-1)\Delta r_1 - \Delta^2(N-1)^2}{N^2 r_1 - \Delta(N-1)^2}. \qquad (5.76)$$

Only the second root is acceptable, and we write it in the form

$$s = \Delta \frac{(N^2-1)}{\alpha N^2}, \qquad (5.77)$$

where

$$\alpha = \frac{1 - \dfrac{\Delta(N-1)^2}{r_1(N^2-1)}}{1 - \dfrac{\Delta(N-1)^2}{N^2 r_1}}. \qquad (5.78)$$

In order to elucidate the behavior of the function in (5.78), we assume that $\Delta = 0.1D$ and that the aperture ratio $\mathbb{A}$ of the primary mirror varies from $1/4$ to $1/3$. Since the ratio f/r_1 lies within the range $(-0.5, -0.3)$ in the application, the quantity

$$x \equiv \frac{\Delta}{r_1} = \frac{\Delta}{D}\frac{D}{f}\frac{f}{r_1} = 0.1\mathbb{A}\frac{f}{r_1}$$

can be found in the interval $(0.007, 0.016)$. It is straightforward to verify, for BK7 glass for instance, that the function (5.78) is actually very close to 1.001. Therefore, the condition for axial achromatism is

$$s = \Delta \frac{(N^2-1)}{N^2}. \qquad (5.79)$$

Although this equation leads to axial achromatism, it is more convenient to use the following relation instead:

$$s = \Delta \frac{(N^2-1)}{\alpha N^2}. \quad \alpha \in (0.95, 1) \qquad (5.80)$$

This is because (as demonstrated by the examples in **MCamera**), for $\alpha = 0.975$, rays of different colors that intercept C_M at $0.7D$ arrive at the same point on the axis. If $\alpha = 0.95D$, this happens to the marginal rays. From now on, we will assume that the thickness $\Delta = 0.1D$ and that the difference $s = r_1 - r_2$ is given by (5.80).

To determine the values of r_1, we now need to impose the condition that the spherical aberration of the whole combination vanishes:

$$D_{S,1} + D'_{S,1} = 0. \tag{5.81}$$

Because of (5.71) and (5.73), this is a seventh-degree algebraic equation in the unknown r_1, which can be solved with **MCamera** when the *numerical* values of N and Δ are given. It is clear that some of these solutions may be meaningless from a physical point of view. In any case, this equation has two real roots with opposite signs. In practice, the negative root is usually chosen.

It is evident that this value of r_1 represents a first approximation since it is obtained by calculating the spherical aberration of the mirror S without taking the presence of the corrector C_M into account. Moreover, no information about the distance δf between C_M and S has been provided so far.

MCamera can supply the values of r_1 and δ that will reduce the third-order spherical aberration and coma to zero. The input data are D, f, and the field angle θ, and we will assume that C_M is made from BK7 glass. This software also evaluates the value of r_2 as well as the other aberrations in the system.

Note that higher-order spherical aberrations are significant when the surfaces of the corrector are highly curved. Therefore, a design based on the third-order theory is not usually acceptable without resorting to some optimization techniques.

Examples are given in the notebook **MCamera**, where it is shown that good results are obtained for speeds of up to $F/4$.

5.9 Examples

The corrector has a diameter of $D = 200$ mm and a thickness of $0.1D = 20$ mm.

Figure 5.18 shows the spot diagram for a Maksutov camera with a primary of speed $f/4$. It is clear that this camera will not be satisfactory for photographic purposes due to its higher-order aberrations. Figure 5.19 shows the spot diagram for a Maksutov camera with a mirror of speed

5.9. Examples   125

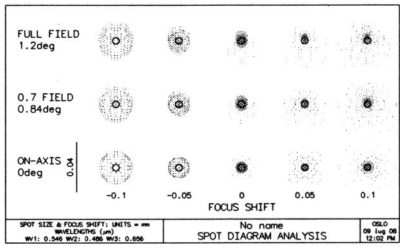

Fig. 5.18 Spot diagram for a Maksutov camera of speed $f/4$

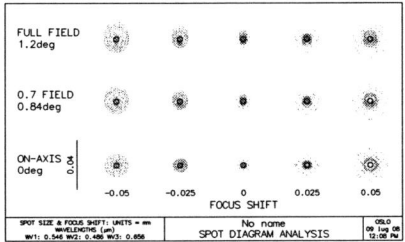

Fig. 5.19 Spot diagram for a Maksutov camera of speed $f/3.3$

$f/3.3$, where the camera was designed with the software package **Optisoft** (see the Preface). It is evident that this faster camera will be good enough if intended for photography.

Chapter 6

Compound Cassegrain Telescopes

6.1 Introduction to Cassegrain Telescopes

Very efficient optical configurations can be obtained for astronomical telescopes by combining one of the correctors described in the previous chapter with the Cassegrain scheme. Depending on the corrector chosen, a Schimdt–Cassegrain, Houghton–Cassegrain or Maksutov–Cassegrain telescope is obtained (see Figure 6.1).

First, note that it is simpler to eliminate the aberrations in one of the aforesaid configurations than in the corresponding photographic camera for the following two reasons:

- Due to the longer focal length of a telescope compared to a camera, the field angle and the off-axis aberrations are smaller

- The secondary mirror partially compensates for the aberrations of the primary mirror.

Thus, the corrector must balance out the residual aberrations of a Cassegrain configuration for small field angles. However, it is important to note that much more accurate correction is needed if we wish to reach the diffraction limit.

This chapter reviews the optical combinations that are obtained by associating a Schmidt corrector C_S, a Houghton corrector C_H, or a Maksutov corrector C_M with the Cassegrain scheme (see Figure 6.1).

The Cassegrain design needs to be developed as shown in Section 4.2: based on the diameter and focal length f_1 of S_1, the total focal length f of the combination, and the distance between S_1 and the focal plane, it is possible to determine the diameter of S_2, its radius of curvature R_2, and the distance $s_{1,2}$ between S_1 and S_2 (back focal), or (equivalently) the nondimensional quantities β and γ.

In the second step, we determine the aberrations of S_1 and S_2 in the absence of the corrector C but the presence of an aperture stop a distance

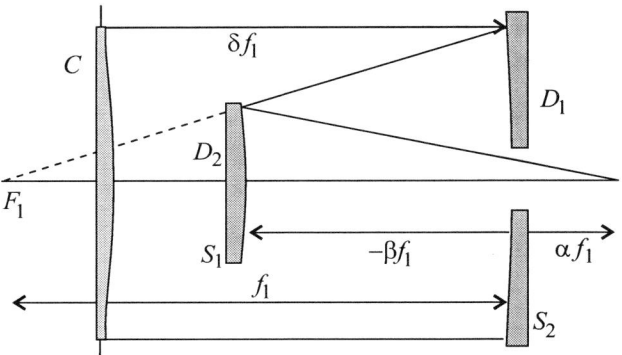

Fig. 6.1 Compound Cassegrain telescope

δf from S_1, proceeding as described in the previous chapter. As already noted, this approach is possible because the corrector is an almost afocal optical element that does not modify the Gaussian characteristics of the whole system (see Section 5.1).

The calculations involved—although simple—are very tedious, so it is again advantageous to use **Mathematica**® in order to quickly obtain the final results. By resorting to the notebook *SCassegrain*, the coefficients D_1, D_2 and D_3 of spherical aberration, coma and astigmatism for the whole system comprising S_1 and S_2 with an aperture stop at a distance δf_1 from S_1 are obtained:

$$32 f_1^3 D_1 = B_{1,0} - K_1 + B_{1,2} K_2,$$
$$8 f_1^2 D_2 = B_{2,0} - \delta K_2 + B_{2,2} K_2, \quad (6.1)$$
$$8 f_1 D_3 = B_{3,0} - \delta^2 K_2 + B_{3,2} K_2$$

where

$$B_{1,0} = -1 + \frac{(\alpha+1)(M^2-1)}{M^3}, \quad B_{1,2} = \frac{(M-1)^3(1+\alpha)}{M^3(1+M)},$$
$$B_{2,0} = \frac{M + \alpha - M^3(\delta-1) - \delta(\alpha+1) + M^2(\alpha(\delta-1)+\delta)}{M^3}, \quad (6.2)$$
$$D_{2,2} = \frac{(M-1)^3(M + \alpha(\delta-1)+\delta)}{M^3(1+M)},$$

$$B_{3,0} = \frac{1}{M^3(1+\alpha)} \left(M^4 + 2M(\alpha(\delta-1)+\delta) - (\alpha(\delta-1)+\delta)^2 + \right.$$
$$\left. M^3(1+\alpha)(\delta-1)(1+\alpha(\delta-1)+\delta) - M^3(4 - 2\delta + \delta^2 + \alpha(2 - 2\delta + \delta^2)) \right),$$
$$(6.3)$$

$$B_{3,2} = \frac{(M-1)^2(M+\alpha(\delta-1)+\delta)^2}{M^3(M+1)(\alpha+1)}. \tag{6.4}$$

6.2 Schmidt–Cassegrain Telescopes

We now have all of the ingredients required to develop a Schmidt–Cassegrain combination. In this case, there are three possibilities:

- Both primary and secondary mirror are spherical and the corrector must eliminate the spherical aberration of the whole combination. This is realized when

$$B_{1,0} + a_4(N-1) = 0, \tag{6.5}$$

that is, when

$$a_4 = \frac{B_{1,0}}{N-1}. \tag{6.6}$$

The coma and astigmatism are acceptable if the speed of S_1 is less than or equal to $F/4$. In any case, coma can be eliminated if the distance δf_1 between the corrector and the primary mirror is equal to the (positive) root of the equation

$$B_{2,0} = 0. \tag{6.7}$$

It is straightforward to verify that this value is

$$\delta = \frac{M(M^2+1) + \alpha(1-M^2)}{1+M^3+\alpha-M^2(1+\alpha)}, \tag{6.8}$$

and that it corresponds to a lengthening of the tube.

- The primary is spherical and the secondary is characterized by a conic constant K_2. In this case, the coefficient a_4 of the corrector and the conic constant K_2 are used to eliminate the spherical aberration and coma. This happens when

$$B_{1,0} + B_{1,2}K_2 + a_4(N-1) = 0, \quad B_{2,0} + B_{2,2}K_2 = 0. \tag{6.9}$$

This system is satisfied for

$$a_4 = \frac{B_{1,2}D_{2,0} - B_{1,0}B_{2,2}}{B_{2,2}(N-1)}, \quad K_2 = -\frac{B_{2,0}}{B_{2,2}}. \tag{6.10}$$

- Finally, the spherical aberration, coma and astigmatism can be eliminated by utilizing the three degrees of freedom a_4, K_1 and K_2 and imposing that they satisfy the system

$$B_{1,0} - \frac{1}{32 f^3} K_1 + B_{1,2} K_2 + a_4 (N-1) = 0,$$

$$B_{2,0} - \frac{1}{8 f^2} \delta K_1 + B_{2,2} K_2 = 0, \qquad (6.11)$$

$$B_{3,0} - \frac{1}{8 f} \delta^2 K_1 + B_{3,2} K_2 = 0.$$

The corresponding solution is

$$a_4 = \frac{1}{4 f_1 (N-1) \delta (D_{2,2} f_1 \delta - D_{3,2})} \left(-B_{2,0} D_{3,2} - 4 B_{1,2} B_{3,0} f_1 \delta + 4 B_{1,0} B_{3,2} f_1 \delta + 4 B_{1,2} B_{2,0} f_1^2 \delta^2 + B_{2,2} (B_{3,0} - 4 B_{2,0} f_1^2 \delta^2) \right),$$

$$K_1 = \frac{8 (B_{2,2} B_{3,0} - B_{2,0} B_{3,2}) f_1^2}{\delta (B_{2,2} f_1 \delta - B_{3,2})},$$

$$K_2 = \frac{B_{3,0} - B_{2,0} f_1 \delta}{B_{2,2} f_1 \delta - B_{3,2}}.$$

(6.12)

For the above cases, note that:

- First, it is essential to eliminate the spherical aberration. However, one or two conic mirrors are needed to eliminate either coma or both coma and astigmatism, and this is very difficult to realize in practice.

- If the primary mirror has a speed of more than $f/3.5$, the curvature of the second surface of the corrector becomes appreciable, so we must also consider the aberrations introduced by it.

- Due to the high speed of the primary mirror, fifth-order spherical aberration occurs, and so that the above values of a_4, K_1 and K_2 do not ensure that the corresponding combination performs well.

In order to tackle the second problem, we should evaluate the small variations in a_4, K_1 and K_2 due to the radius of curvature R_c. In regards to the last problem, there are two ways to eliminate fifth-order spherical aberration:

- We could consider the sixth-order profile of the corrector

$$y = a_2 y^2 + a_4 y^4 + a_6 y^6,$$

and try to work out which value of a_6 eliminates the fifth-order spherical aberration. However, such a corrector is very difficult to produce.

6.3. Examples

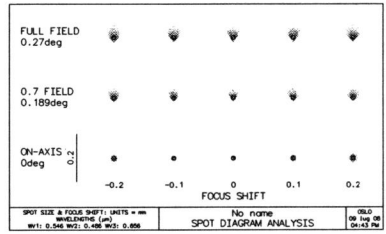

Fig. 6.2 Schmidt–Cassegrain with spherical mirrors

- Alternatively, at least for primary mirrors that are slower than $f/2.5$, we could try to balance out the third- and fifth-order spherical aberrations. This approach requires deeper analysis of the optical combination.

6.3 Examples

In this section we provide examples of Schmidt–Cassegrain telescopes obtained using the notebook *SCassegrain*. For all of the designs considered here, the diameter of the principal mirror S is 200 mm. In the first three examples, the primary mirror has a speed of $f/3$; in the final three examples, the primary mirror has a speed of $f/2$.

First, suppose that S has a speed $f/3$ and that we are searching for the data on a telescope with a total focal length of 2400 mm, a back focal of 250 mm, and where the corrector and primary mirror are separated by 450 mm. If we denote the radius of the secondary mirror S_s by R_s, the distance between S and S_s by δf, the radius of the second surface of the corrector C at its vertex by R_C, the aspheric constant of C by a_4, and the conic constants of S and S_s by K_1 and K_2, respectively, then, for the three cases considered in the previous section, the notebook *SCassegrain* gives

$$R_s = -453.33, \quad \delta f = -430.$$

- Third-order spherical aberration is absent when:

$$R_c = -715725, \quad a_4 = 1.863 \times 10^{-10}.$$

- Third-order spherical aberration and coma are absent when

$$R_c = -590650, \quad a_4 = 2.257 \times 10^{-10}, \quad K_2 = -1.18.$$

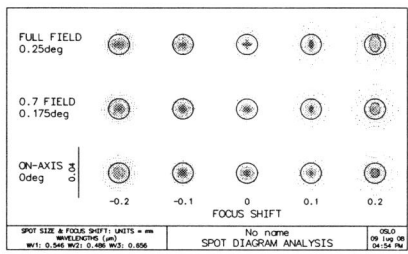

Fig. 6.3 Schmidt–Cassegrain with conic secondary

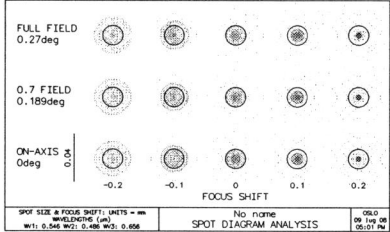

Fig. 6.4 Schmidt–Cassegrain with conic mirrors
(focal surface radius: −380 mm)

- Third-order spherical aberration, coma, and astigmatism are absent when

$$R_c = -365946, \quad a_4 = 3.644 \times 10^{-10}, \quad K_1 = 0.644, \quad K_2 = 0.005.$$

Figures 6.2, 6.3, and 6.4 show spot diagrams corresponding to the above data on focal spherical surfaces of radius −380 mm. We can see that, in the third case, higher-order aberrations reduce the performance of the combination.

In order to shorten the tube containing the optical combination, we now consider a primary mirror S with a speed of $f/2$ and a diameter of 200 mm. The new input data are: total focal length 2000 mm; back focal 200 mm; distance between corrector and primary mirror 305 mm. The notebook *SCassegrain* then gives

$$R_s = -250, \quad \delta f = -300.$$

6.4. Houghton–Cassegrain Telescopes   133

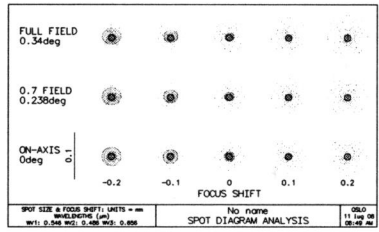

Fig. 6.5 Schmidt–Cassegrain with conic mirrors
(focal surface radius: −200 mm)

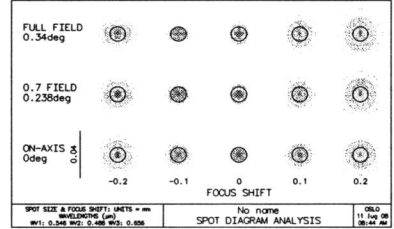

Fig. 6.6 Schmidt–Cassegrain with conic mirrors
(focal surface radius: −380 mm; designed with *Optisoft*)

Third-order spherical aberration and coma are absent when

$$R_c = -173043, \quad a_4 = 7.705 \times 10^{-10}, \quad K_2 = -0.833.$$

The spot diagram on a focal surface of radius −200 mm shown in Figure 6.5 indicates that third-order correction is not sufficient in this case.

Figure 6.6 shows a spot diagram for the same combination designed with *Optisoft*.

6.4 Houghton–Cassegrain Telescopes

An aplanatic telescope with **spherical surfaces** is obtained by combining the Cassegrain scheme with Houghton's corrector (see Figure 6.7).

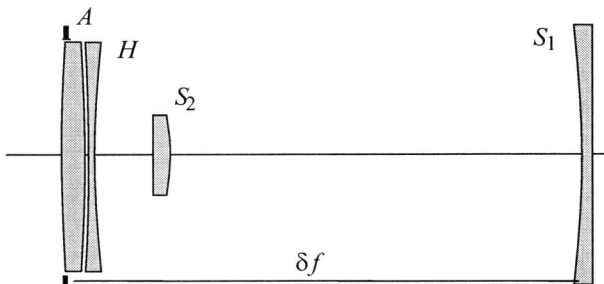

Fig. 6.7 Houghton–Cassegrain telescope

All of the results derived in Chapter 5 for Houghton cameras can be extended to this case. In particular, for a symmetric doublet we have

$$R_1 = -\frac{2(N^2-1)D_1D_2}{2(N+1)D_1^2 + N(N-1)D_2^3},$$
$$R_2 = \frac{2(N^2-1)D_1D_2}{2(N+1)D_1^2 - N(N-1)D_2^3}, \quad (6.13)$$

where the aberration coefficients D_1 and D_2 are given by (6.1) and (6.2) for $K_1 = K_2 = 0$.

For a fixed pair of mirrors S_1 and S_2 in a Cassegrain combination, the radii R_1 and R_2 depend on the magnification M and the distance factor δ. Figures 6.8 and 6.9 (see the notebook **HCassegrain**) show plots of the nondimensional quantities

$$r_1 = \frac{R_1}{f_1}, \quad r_2 = \frac{R_2}{f_1}$$

versus δ for $M = 3$ and $M = 4$, respectively. In both plots, the dotted line refers to r_2. It is evident that the most convenient values of r_1 and r_2 (i.e., the largest absolute values that reduce higher-order aberrations) are obtained for $\delta \in [0.7, 1]$. On the other hand, we must increase the magnification M if we wish to reduce the obstruction factor. However, the plots in Figure 6.8 show that the radii r_1 and r_2 decrease and higher-order aberrations become more relevant as M increases.

According to the above formulae, very good results are obtained when the principal mirror S is slower than $f/3.5$. When the speed of S is faster than $f/3.5$ and slower or equal to $f/3$, the curvatures supplied by (6.13) are too strong and the higher-order aberrations dominate.

One way to reduce the curvature of the lenses, and consequently the higher-order aberrations, is to use *four* different values for the radii of the

6.4. Houghton–Cassegrain Telescopes

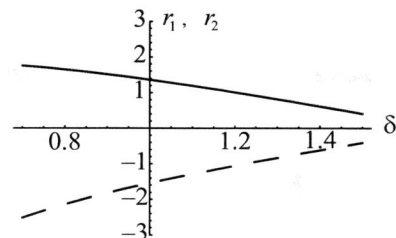

Fig. 6.8 f_1, r_2 versus δ, for $M = 3$

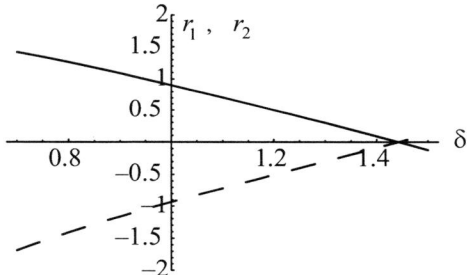

Fig. 6.9 f_1, r_2 versus δ, for $M = 4$

lenses comprising the doublet. Again, all of the results in Chapter 5 can be applied to this new situation.

First, we need to determine the region where the discriminant $\Delta > 0$ upon varying δ and r_3 (see Figure 6.10). Figure 6.5 shows this region for BK7 glass and $\alpha = 0.4$, where α is the ratio between the back distance and the primary focal, and for a magnification factor $M = 4$.

Moreover, using **Mathematica**® it is possible to analyze the behavior of the radii r_1 and r_2 as r_3 is varied in the set I_δ of admissible values corresponding to a given δ (see Chapter 5). Figures 6.11–6.12 plot the variables r_1, r_2, and r_4 versus r_3 for $\delta = 0.8$. More precisely, the first figure refers to solution (5.64) and the second one to (5.65). The interval chosen for δ starts from the lowest bound of I_δ (see **HCassegrain**).

In all of the figures shown in this section, the continuous line refers to r_1, the line with longer dashes to r_2, and the line with shorter dashes to r_4.

Finally, Figures 6.13–6.14 are similar to Figures 6.11–6.12 but show the results for $\delta = 1.1$.

Careful analysis of the plots leads us to the following conclusions:

- The highest and thus most convenient values of r_1, r_2, r_3 and r_4 are obtained for $\delta \in [0.7, 1]$.

136   Chapter 6. Compound Cassegrain Telescopes

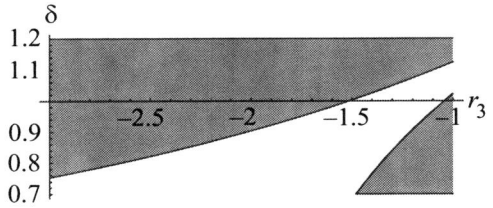

Fig. 6.10 The region in which $\Delta > 0$, for $M = 4$

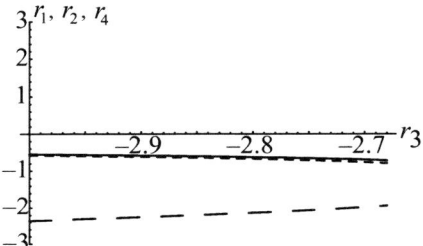

Fig. 6.11 Plots of r_1, r_2, and r_4 versus r_3 for $\delta = 0.8$ and solution (5.64)

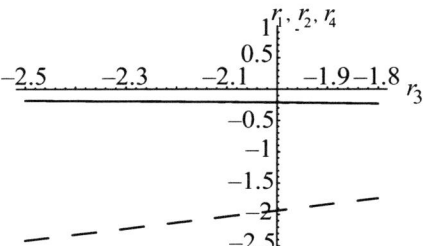

Fig. 6.12 Plots of r_1, r_2, and r_4 versus r_3 for $\delta = 0.8$ and solution (5.65)

- For an assigned δ in this interval, solution (5.64) and r_3, r_4 attain their highest values at the lowest bound of I_δ, whereas solution (5.65) and r_3, r_4 must be evaluated at 0.8–0.9 times the lowest bound. For instance, for $\delta = 0.8$, we derive $I_\delta = [-1.65, -0.8]$ from Figure 6.3, meaning that solution (5.64) must be evaluated at $r_3 = -1.65$, whereas solution (5.65) should be evaluated at $r_3 = -1.5$ (see

6.5. Examples

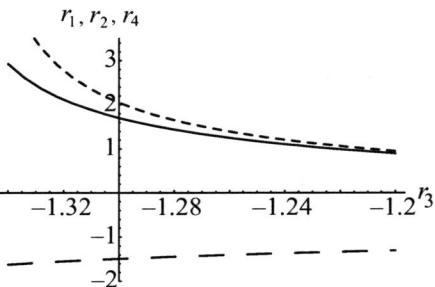

Fig. 6.13 Plots of r_1, r_2, and r_4 versus r_3 for $\delta = 1.1$ and solution (5.64)

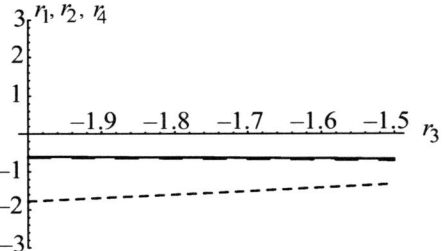

Fig. 6.14 Plots of r_1, r_2, and r_4 versus r_3 for $\delta = 1.1$ and solution (5.65)

Figures 6.13 and 6.14, which refer to solutions (5.64) and (5.65), respectively).

- In any case, the second solution leads to larger radii.

6.5 Examples

In this section we provide some examples of Houghton–Cassegrain telescopes. We start with a principal mirror S of diameter $D = 200$ mm and speed $f/3$. We denote the focal length of the primary mirror by f_1, the total focal length of the whole combination by f_t, the thicknesses of the doublet by $spes$, the distance between the corrector and S by δf_1, the back focal by em, the refractive index of the glass of the doublet by N_i, the view angle by θ, the radius of curvature of the secondary mirror by R_s, the

138   Chapter 6. Compound Cassegrain Telescopes

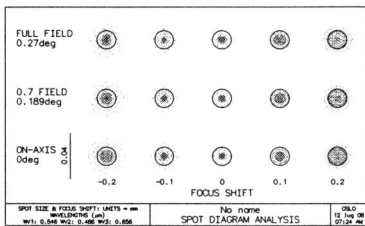

Fig. 6.15 Houghton–Cassegrain; speed of primary: $f/3$

distance between the primary and secondary mirrors by Δ, and the radii of curvature of the surfaces of the doublet by R_i, $i = 1, \ldots, 4$. The input data for the program ***HCassegrain2*** are then:

$$f_1 = 600$$
$$f_t = 2400$$
$$spes = 15, 5, 5$$
$$em = 200$$
$$D = 200$$
$$\delta = 480/600$$
$$N_i = 1.5187228 \text{ (BK7 in green light)}$$
$$\theta = 0.27.$$

The corresponding output is:

$$R_s = -426,667$$
$$\Delta = -440$$
$$R_1 = 53898$$
$$R_2 = -1168$$
$$R_3 = -795.5$$
$$R_4 = -2616.$$

Figure 6.15 shows the spot diagram for this combination, which is diffraction limited.

Now we will attempt to design a similar combination with a shorter tube and a higher speed. To this end we start with a faster primary mirror S, of speed $f/2.25$, and we consider the following input data:

$$f_1 = 450$$
$$f_t = 2000$$
$$spes = 10, 10, 5$$
$$em = 200$$
$$D = 200$$
$$\delta = 340/450$$
$$N_i = 1.5187228 \text{ (BK7 in green light)}$$
$$\theta = 0.3.$$

6.5. Examples

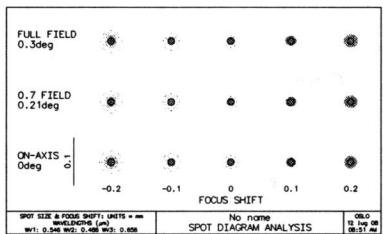

Fig. 6.16 Houghton–Cassegrain, speed of primary: $f/2.25$

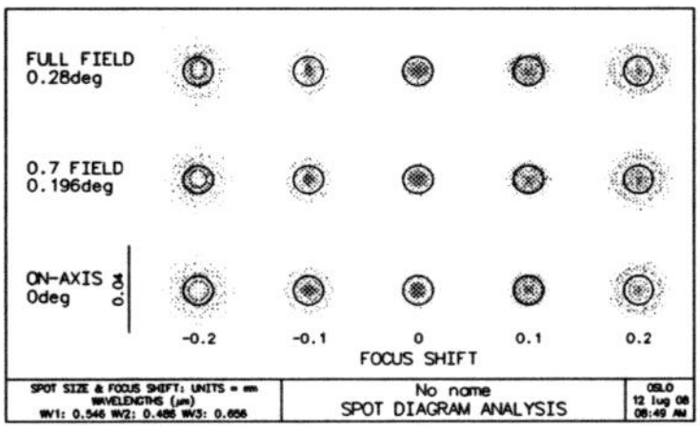

Fig. 6.17 Houghton–Cassegrain combination (speed of primary: $f/2.25$) designed using *Optisoft*

The corresponding output data obtained with the notebook **HCassegrain2** are listed below:

$$R_s = -308.1$$
$$\Delta = -330.6$$
$$R_1 = 84784$$
$$R_2 = -868.1$$
$$R_3 = -593.04$$
$$R_4 = -1914.$$

The spot diagram for this combination (see Figure 6.16), which is produced on a focal surface of radius -200 mm, shows that the third-order correction is not acceptable due to the presence of significant higher-order aberrations introduced by the shorter radius of curvature of the primary. For instance, we must balance out the third-order and fifth-order aberrations, improve the spherochromatism, etc. In other words, increasing the speed of the primary does not improve the third-order design and so we need to analyze the problem more accurately. For instance, Figure 6.17 shows the spot diagram for a Houghton–Cassegrain combination that was designed using the software *Optisoft* and the above data. It is clear that this combination is again diffraction limited.

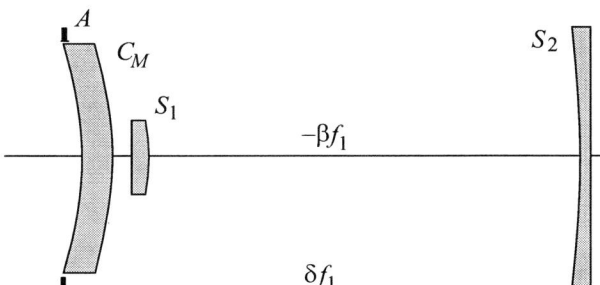

Fig. 6.18 Maksutov–Cassegrain telescope

6.6 Maksutov–Cassegrain Telescopes

A Maksutov–Cassegrain telescope is obtained by placing a Maksutov corrector C_M in front of a Cassegrain combination (see Figure 6.18).

Although there are fewer surfaces in this scheme than in the corresponding Houghton–Cassegrain telescope, this combination is much more difficult to analyze, for the following reasons:

- The thickness of the meniscus can not be neglected, even to a first approximation, since it plays a fundamental role in compensating for the aberrations.

- The focal length of the meniscus, although still much greater than the focal length of the primary mirror, modifies the Gaussian behavior of the Cassegrain part. In other words, the principle that was stated at the beginning of Chapter 5 and has been applied in all previous designs leads to unacceptable designs in this case.

Bearing these remarks in mind, we apply the same procedure that we used in previous sections. If f_1 is the focal length of the primary mirror S, γf_1 is the focal length of the secondary mirror S_s, em is the back focal, f_t is the total focal length of the Cassegrain combination, δf_1 is the distance between the Maksutov corrector M and S, and $-\beta f_1$ is the distance between S and S_s, then the coefficient of spherical aberration for the Cassegrain combination where the aperture stop is located δf from the primary mirror S is (see Chapter 4 or *MCassegrain*):

$$D_{S,1} = -\frac{1 + M^3 + \alpha - M^2(1+\alpha)}{32 f_1^3 M^3} \qquad (6.14)$$

6.6. Maksutov–Cassegrain Telescopes

where

$$\alpha = \frac{em}{f_1}, \tag{6.15}$$

$$M = \frac{f_t}{f_1}, \tag{6.16}$$

$$\beta = \frac{M-\alpha}{M+1}, \tag{6.17}$$

$$\gamma = \frac{2M(1+\alpha)}{1-M^2}. \tag{6.18}$$

Moreover, if R_1 and $R_1 - s$ are the radii of the two surfaces of M, then the coefficient of spherical aberration for M is obtained (see Section 5.8 and **MCassegrain**):

$$D_{M,1} = \frac{N-1}{8N^3 R_1^4 (R_1-s)^3}[(R_1 + (N-1)(s-\Delta))^2(N(R_1-\Delta)+\Delta)$$
$$(R_1 + (N-1)(s+Ns-N\Delta) - R_1 N(R_1-s)^3], \tag{6.19}$$

where, in order to get an achromatic corrector, we must take

$$s = \Delta \frac{N^2-1}{0.97 N^2}. \tag{6.20}$$

Finally, we choose $\Delta = 0.1D$, where D is the diameter of M.

When a Cassegrain combination is employed, an approximate condition for eliminating spherical aberration is

$$D_{s,1} + D_{M,1} = 0. \tag{6.21}$$

As shown in **MCassegrain** and Chapter 5, (6.21) is a seventh degree algebraic equation that yields only three real roots. One of them is too small to be useful. The remaining two roots have opposite signs; we choose the negative root.

Considering the remarks made at the beginning of this chapter, (6.21) gives unacceptable results for two main reasons:

- The strong curvatures of the surfaces of M introduce higher-order spherical aberration

- The high but finite negative focal length of M significantly modifies the focal length of the Cassegrain combination.

The next section provides some advice on overcoming these problems.

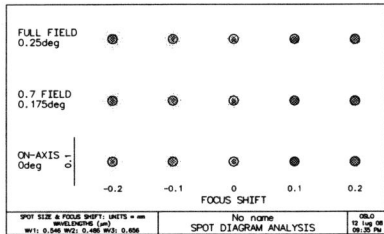

Fig. 6.19 Spot diagram for a Maksutov–Cassegrain combination with a primary of speed $f/3.5$

6.7 Examples

The notebook *MCassegrain* allows us to design a Maksutov–Cassegrain telescope with a speed of $f/11-f/14$. To obtain acceptable results, we start with a focal length f_c of the Cassegrain combination that is less than the desired total focal length f_t. This strategy is employed because of the effect of the corrector, which increases the focal length of the Cassegrain combination. We can find the correct initial value of f_c after a few attempts. Moreover, the radius of the second surface of M, which is obtained using the notebook *MCassegrain*, must be reduced slightly in order to introduce a third-order spherical aberration that can balance out the fifth-order one. The following examples illustrate this procedure.

We wish to design a Maksutov–Cassegrain combination with a focal length of 2800 mm, starting with a principal mirror S of diameter $D = 200$ mm and speed $f/3.5$. We introduce the following input data:

$$f_1 = 700$$
$$f_t = 2400$$
$$em = 200$$
$$\delta = 520/700$$
$$D = 200$$
$$\theta = 0.25,$$

where θ is the field angle. The corresponding output is:

$$R_1 = -300.06$$
$$R_2 = -311.74$$
$$-\beta f_1 = -496.77$$
$$R_s = -573.814.$$

6.7. Examples   143

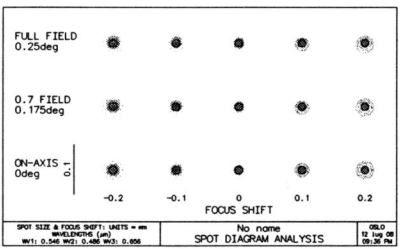

Fig. 6.20 Spot diagram for a Maksutov–Cassegrain combination with a primary of speed $f/3$

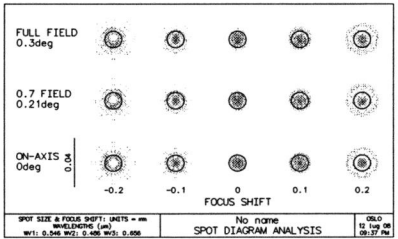

Fig. 6.21 Spot diagram for a Maksutov–Cassegrain combination with primary of speed f/2.7 designed using *Optisoft*

Although third-order spherical aberration is absent from this combination, the spot diagram shows that it is still not acceptable. Therefore, we reduce the value of R_2 to -311.45 step by step. The corresponding spot diagram is given in Figure 6.19.

Similarly, to design a Maksutov–Cassegrain telescope with a focal length $f_t = 2200$ mm and a principal mirror of speed $f/3$, we introduce the following input data:
$$f_1 = 600$$
$$f_t = 1950$$
$$em = 200$$
$$\delta = 450/600$$
$$D = 200$$
$$\theta = 0.5.$$

The corresponding output is:
$$R_1 = -269.5$$
$$R_2 = -281.18$$
$$-\beta f_1 = -411.76$$
$$R_s = -543.79.$$

Again, the spot diagram is not acceptable and so we reduce R_2 to the value -280.72 step by step. This value yields a good result, as shown by the spot diagram in Figure 6.20.

This approach does not work for higher aperture ratios. Figure 6.21 shows the spot diagram for a Maksutov–Cassegrain telescope with a focal length of 2000 mm and a primary mirror of speed $f/2.7$ that was designed using ***Optisoft***.

6.8 Final Remarks

In this chapter, and in the previous one, astronomical combinations were presented that are essentially based on mirrors, and more specifically on the Cassegrain combination. The dioptric elements (Schmidt, Houghton or Maksutov corrector) modify the Gaussian performance of the Cassegrain part as little as possible and are employed in order to balance out some of the third-order aberrations without introducing significant chromatism.

Advantages of these combinations include the following:

- They are very compact
- For any mirror, there is only one surface to consider
- The homogeneity and the refractive index of the glass of the mirror are not important
- Large diameters can be realized at reasonable cost.

These advantages are counterbalanced by the following negative aspects:

- They exhibit sensitive turbulence phenomena (especially the open Cassegrain combination)
- The obstruction caused by the secondary mirror modifies the diffraction pattern, reducing the resolving power of the instrument.

Chapter 7

Doublets and Triplets

7.1 Achromatic Doublets

An achromatic doublet is an optical system S of two lenses L_1 and L_2 with different refractive indices and Abbe numbers (see (3.106)) that are separated by a very thin layer of air. In the **Fraunhofer doublet**, the first lens is positive and the second lens is negative, whereas the opposite is true in the **Steinheil doublet**. Moreover, the positive lens has a lower refractive index than the negative lens, but its Abbe number is much greater than that of the negative lens. The Steinheil design is used when it offers a better correction than the Fraunhofer or when the positive element exhibits poor resistance to weathering.

More precisely, a doublet is said to be **achromatic** if the following conditions are satisfied:

1. The focal f has a given value

2. The back focal of blue light bf_b and the back focal of red light bf_r are equal[1]

3. The third-order spherical aberration in green light vanishes

4. The third-order coma in green light vanishes.

In order to satisfy the above conditions, we manipulate the type of glass used, the thickness of the lenses and their radii of curvature. In particular, the four radii of the lenses can be chosen to satisfy the above four conditions. However, the algebraic system A that is obtained by expressing these four conditions analytically is so complex that it is impossible to solve it, even using a powerful program like ***Mathematica***®. Consequently, we instead

---

[1] Here and in what follows, green light corresponds to the line e (546.07 nm), blue light to the line F (486.13 nm), and red light to the C line (656.27 nm).

try to determine approximate values for the roots of A. This is usually done by ignoring the thickness of the lenses (i.e., by assuming that the doublet consists of thin lenses). According to this hypothesis, if we denote the surfaces of the lenses L_1 and L_2 by S_i, $i = 1, \ldots, 4$ and the curvature of the surface S_i by $c_i = 1/r_i$, and we introduce the **bendings** of the two lenses $\xi_1 = c_1 - c_2$ and $\xi_2 = c_3 - c_4$, then it is well known that the first two conditions are expressed by the system (see the notebook **Doublet**)

$$(N_1 - 1)\xi_1 + (N_2 - 1)\xi_2 = 1/f, \tag{7.1}$$
$$(N_{1,b} - N_{1,r})\xi_1 + (N_{2,b} - N_{2,r})\xi_2 = 0. \tag{7.2}$$

In the above formulae N_i, $N_{i,b}$, and $N_{i,r}$ denote the refractive indices of lens L_i in green light, blue light, and red light, respectively. If we refer all of the lengths that appear in (7.1) and (7.2) to the focal f and denote the nondimensional quantities with the same symbols, then the above system can be written in the form

$$\alpha_1 V_1 \xi_1 + \alpha_2 V_2 \xi_2 = 1, \tag{7.3}$$
$$\alpha_1 \xi_1 + \alpha_2 \xi_2 = 0, \tag{7.4}$$

where we have introduced the Abbe number

$$V_i = \frac{N_i - 1}{N_{i,b} - N_{i,r}} \tag{7.5}$$

of L_i and the notation

$$\alpha_i = N_{i,b} - N_{i,r}. \tag{7.6}$$

It is straightforward to verify that the solutions of (7.3) and (7.4) are

$$\xi_1 = c_1 - c_2 = -\frac{1}{\alpha_1(V_2 - V1)} \tag{7.7}$$
$$\xi_2 = c_3 - c_4 = \frac{1}{\alpha_2(V_2 - V1)}, \tag{7.8}$$

provided that the Abbe numbers of the two glasses are not equal. Thus, we need to construct the two lenses from glass with different Abbe numbers in order to find the possible values of the bendings ξ_1 and ξ_2 for the two lenses. Since $\alpha_i \in (0.005, 0.02)$ for real glasses, we will obtain small differences in curvature *if the difference $V_2 - V_1$ is as large as possible.* We conclude by noting that, upon imposing the first two conditions that define an achromatic doublet, we can express c_1 in terms of c_2 and c_3 in terms of c_4 using (7.7) and (7.8).

7.2 Elimination of Spherical Aberration and Coma

It remains for us to determine the conditions under which the third-order spherical aberration and coma vanish. The calculations are not easy for thin lenses. However, using the notebook *TotalAberrations* and taking into account (7.7) and (7.8), we find that coma vanishes when

$$N_2(N_1^2 - 1)(V_2 - V_1)\alpha_1\alpha_2^2 c_2 + N_1(N_2^2 - 1)(V_1 - V_2)\alpha_1^2\alpha_2 c_4$$
$$+ N_1(1 - 2N_2^2 + N_2^3)\alpha_1^2 + N_1(N_1 - 1)(1 + N_2 - 2N_2^2)\alpha_1\alpha_2$$
$$+ N_2(1 - 2N_1^2 + N_1^3)\alpha_2^2 = 0. \tag{7.9}$$

Finally, the spherical aberration vanishes when the following quadratic condition between c_2 and c_4 is satisfied:

$$(2 - N_1 - N_1^2)N_2(V_1 - V_2)^2\alpha_1^2\alpha_2^3 c_2^2$$
$$-(2 - N_2 - N_2^2)N_1(V_1 - V_2)^2\alpha_1^3\alpha_2^2 c_4^2$$
$$+(4 - 3N_1 - 3N_1^2 + 2N_1^3)N_2(V_1 - V_2)\alpha_1\alpha_2^3 c_2$$
$$[N_1(N_2 - 1)(V_1 - V_2)\alpha_1^2\alpha_2 \times$$
$$((-4 - N_2 + 2N_2^2)\alpha_1 - 4(N_1 - 1)(1 + N_2)\alpha_2)]c_4$$
$$N_1(-2 + 2N_2 + 2N_2^2 - 3N_2^3 + N_2^4)\alpha_1^3$$
$$-(N_1 - 1)N_1(4 - N_2 - 6N_2^2 + 3N_2^3)\alpha_1^2\alpha_2$$
$$+(N_1 - 1)^2 N_1(-2 - N_2 + 3N_2^2)\alpha_1\alpha_2^2$$
$$-(-2 + 2N_1 + 2N_1^2 - 3N_1^3 + N_1^4)N_2\alpha_2^3. \tag{7.10}$$

It is evident that the two solutions (c_2, c_4) of the system represented by (7.9) and (7.10) are obtained from very cumbersome expressions. We therefore assign the types of of glass BAK1 and FK54 to the system and analyze it. For these types of glass, (7.7), (7.8), (7.9), and (7.10) become

$$c_1 - c_2 + 3.03207 = 0, \tag{7.11}$$
$$c_3 - c_4 - 6.26055 = 0, \tag{7.12}$$
$$1.620 c_2 - 2.647 c_4 - 10 = 0, \tag{7.13}$$
$$-3.5754 c_2^2 + 5.933 c_4^2 + 1.9026 c_2 + 43.33 c_4 + 120.5 = 0. \tag{7.14}$$

Equations 7.11 and 7.12, which are equivalent to (7.7) and (7.8) respectively, impose the assigned focal and axial chromatism. The remaining two equations, equivalent to (7.9) and (7.10), ensure that the third-order spherical aberration and coma in green light are zero. The solutions (in focal units) for the above system are:

$$(c_2 = -5.2068,\ c_4 = -6.9730),\quad (c_2 = 5.888,\ c_4 = -0.18336). \tag{7.15}$$

148   Chapter 7. Doublets and Triplets

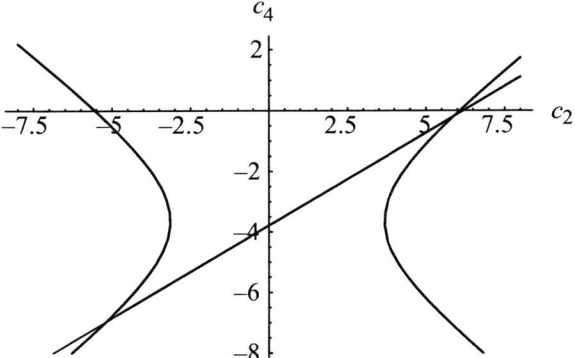

Fig. 7.1 The values of (c_2, c_4) that eliminate spherical aberration and coma are given by the coordinates of the points of intersection of the two curves

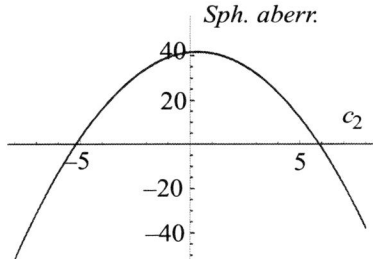

Fig. 7.2 Spherical aberration versus c_2

We can better appreciate these values by plotting the curves that are implicitly defined by equations (7.13) and (7.14) (see Figure 7.1). Moreover, if we introduce the value of c_4 obtained from (7.13) into (7.14), we can derive the relationship between the spherical aberration and the curvature c_2 (see Figure 7.2).

The design based on the above values or obtained using the notebook **Doublet** is usually not acceptable, especially if the speed of the doublet is greater than $f/11$–$f/12$. This is due to the following reasons:

- The thickness of the lenses introduces significant levels of chromatic aberration, coma, spherical aberration and spherochromatism

- The fifth-order aberrations can be significant

- We have no free parameter in the optical system to control the **secondary spectrum**

$$\Delta = bf_b - bf_g, \tag{7.16}$$

where bf_g is the back focal corresponding to green light.

A doublet is said to be **achromatic** if $\Delta/f \simeq 1/2000$, **semiapochromatic** if $\Delta/f \simeq 1/6000$, and **apochromatic** when $\Delta/f \simeq 1/10000$. Now, it is easy to prove that (see [10], p. 311) we have

$$\frac{\Delta}{f} = \frac{P_1 - P_2}{V_1 - V_2}, \tag{7.17}$$

for a thin doublet, where

$$P_i = \frac{N_{b,i} - N_i}{N_{b,i} - N_{r,i}} \tag{7.18}$$

is the dispersion coefficient of the lens L_i. Equation 7.17 implies that the two glasses must be chosen in such a way that *the difference $P_1 - P_2$ is as small as possible and the difference $V_1 - V_2$ is as large as possible.*

Assuming that the types of glass are chosen according to this criterion, then we must compensate for the aberrations introduced by the thickness of the lenses. We can do this either by resorting to an optimization method included in professional software or by carrying out a deeper analysis of the optical system (see ***Optisoft***).

7.3 Examples

In this section we consider some examples of doublets of diameter $D = 100$ mm that were obtained using the notebook ***Doublets***. The input data for this notebook are the set *thick* of lens thicknesses, the set *ind* of refractive indices for green, red and blue light, the diameter D, the field angle θ, and the focal length f. The first of these examples, which is a Fraunhofer objective realized with standard crown and flint glass (BK7-F3), has a speed of $f/15$. The input data are:

$$thick = \{10, 0.5, 7\}$$
$$ind = \{\{1, 1.51872218, 1, 1.61685, 1\}$$
$$D = 100$$
$$\theta = 0.5$$
$$f = 1500.$$

150   Chapter 7. Doublets and Triplets

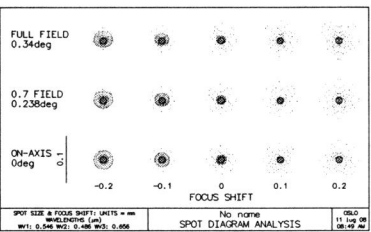

Fig. 7.3 Spot diagram for the BK7-F3 doublet

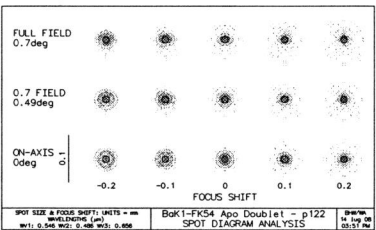

Fig. 7.4 Spot diagram for the BaK1-FK54 doublet

The corresponding output gives the radii of the doublet:

$$\{911.428, -511.906, -519.007, -2261.69\}.$$

Figure 7.3 shows the spot diagram for this combination.
   The second doublet uses BAK1 and F54 glass, and its input data are:

$$thick = \{7, 0.5, 10\}$$
$$ind = \{\{1, 1.5725001, 1, 1.43700005, 1\}$$
$$\{1, 1.57943481, 1, 1.44033746, 1\}$$
$$\{1, 1.56948667, 1, 1.43551944, 1\}\}$$
$$D = 100$$
$$\theta = 0.5$$
$$f = 1000.$$

The corresponding output gives the following radii for the doublet:

$$\{350.148, 169.837, 164.55, -5453.73\}.$$

The spot diagram shown in Figure 7.4 for this doublet shows that the third-order design is not acceptable, since the higher-order aberrations produce a spot that is larger than the Airy disk.

7.3. Examples   151

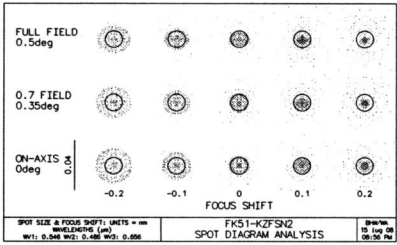

Fig. 7.5 Spot diagram for the Fk51-KZFSN2 doublet

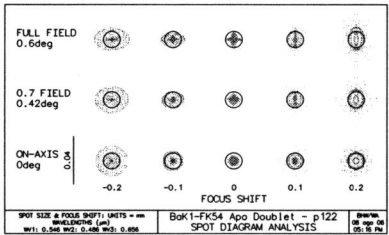

Fig. 7.6 Spot diagram for the Bak1-Fk54 doublet obtained using *Optisoft*

In the third example, we use Fk51 and KZFSN2 glass, and the input data are:

$$thick = \{0, 0, 11\}$$
$$ind = \{\{1, 1.56081792, 1, 1.4879367, 1\}$$
$$\{1, 1.5655173, 1, 1.49055758, 1\}$$
$$\{1, 1.55520714, 1, 1.4847972, 1\}\}$$
$$D = 100$$
$$\theta = 0.5$$
$$f = 1000.$$

The corresponding output yields the following doublet radii:

$$\{422.856, 179.702, 177.399, -11078.1\}.$$

Again, the spot diagram for this combination (see Figure 7.4) shows that the third-order design is not acceptable because the higher-order aberrations produce a spot that is larger than the Airy disk.

Figures 7.6 and 7.7 show the spot diagrams obtained for the latter two combinations when they were redesigned using the software *Optisoft*. Both combinations are now diffraction limited.

152   Chapter 7. Doublets and Triplets

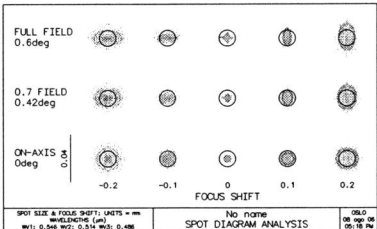

Fig. 7.7 Spot diagram for the Fk51-KZFSN2 doublet obtained using *Optisoft*

7.4 Triplets

A triplet is an optical system S that consists of three lenses L_1, L_2, L_3, each made from different glass. One of the main difficulties encountered when designing a triplet is appropriate glass selection. One criterion that can be applied when making this choice is the following: *in the N_d–V_d diagram, the area of the triangle formed from the points representing the three glasses should be as large as possible* (see [10]).

When the choice of the lenses has been made, we can then consider the thickness of the lenses as well as the radii of curvature of the spherical surfaces of the three lenses. In order to evaluate the approximate values of these quantities, as usual we ignore the thickness of the lenses. Moreover, we analyze a triplet S in which the final two lenses are stuck together. This implies that we only need to determine five radii of curvature R_i, $i = 1, \ldots, 5$. The final design will be acceptable if these radii are large enough to make the higher-order aberrations insignificant.

In order to determine the five radii R_i, $i = 1, \ldots, 5$, we impose the following five conditions:

- The focal length f of S is given;

- The focal lengths in blue light and green light, f_b and f_g respectively, are equal

- The focal lengths in red light and green light, f_r and f_g respectively, are equal

- Third-order coma is absent (in green light)

- Third-order spherical aberration is absent (in green light).

7.4. Triplets

We denote the refractive indices of the three lenses L_1, L_2, L_3 of S, corresponding to green light, blue light, and red light, by N_i, $N_{b,i}$, and $N_{r,i}$ ($i = 1, 2, 3$), respectively. Moreover, we introduce the curvature $c_i = 1/R_i$ of L_i. It is a tedious but simple exercise to verify that the focal length f in green light of the system S is given by the formula

$$(N_1 - 1)(c_1 - c_2) + (N_2 - 1)(c_3 - c_4) + (N_2 - 1)(c_4 - c_5)(N_3 - 1) = \frac{1}{f}. \quad (7.19)$$

Alternatively, (7.19) can be obtained using the notebook **Triplet**. Similarly, the conditions $f_b - f_g = 0$ and $f_r - f_g = 0$ can be written as follows:

$$(N_{1,b} - N_1)(c_1 - c_2) + (N_{2,b} - N_2)(c_3 - c_4)$$
$$+ (N_{3,b} - N_3)(c_4 - c_5) = 0, \quad (7.20)$$
$$(N_{1,r} - N_1)(c_1 - c_2) + (N_{2,r} - N_2)(c_3 - c_4)$$
$$+ (N_{3,r} - N_3)(c_4 - c_5) = 0. \quad (7.21)$$

Equations 7.19–7.21 form a system of three equations in the three unknowns $c_1 - c_2$, $c_3 - c_4$, and $c_4 - c_5$. If we use f as the length unit and introduce the notation

$$\alpha_i = N_i - 1, \quad \beta_i = N_{i,b} - N_i, \quad \gamma_i = N_{r,i} - N_i, \quad (7.22)$$

$$\xi_1 = c_2 - c_3, \quad \xi_2 = c_3 - c_4, \quad \xi_3 = c_4 - c_5, \quad (7.23)$$

the above system assumes the form

$$\alpha_1 \xi_1 + \alpha_2 \xi_2 + \alpha_3 \xi_3 = 0, \quad (7.24)$$
$$\beta_1 \xi_1 + \beta_2 \xi_2 + \beta_3 \xi_3 = 0, \quad (7.25)$$
$$\gamma_1 \xi_1 + \gamma_2 \xi_2 + \gamma_3 \xi_3 = 0, \quad (7.26)$$

whose solutions are (see **Triplet**)

$$\xi_i = \frac{(\beta \times \gamma)_i}{\alpha \cdot \beta \times \gamma}, \quad (7.27)$$

where $\alpha = (\alpha_i)$, $\beta = (\beta_i)$, $\gamma = (\gamma_i)$. Starting from these values of ξ_1, ξ_2, and ξ_3, we can express c_1, c_3, and c_5 as functions of c_2 and c_4. Substituting these relations into the condition that expresses the elimination of third-order coma, we again obtain a *linear* expression containing c_2 and c_4. Finally, the condition for eliminating the third-order spherical aberration becomes a quadratic relation in c_2. The most convenient solution corresponds to the one with the lowest curvatures c_i. All of the above calculations can be carried out using **Triplet**.

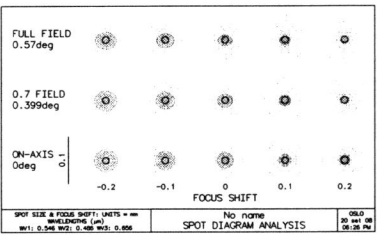

Fig. 7.8 Spot diagram for the Bk7-KZFS1-BaF4 triplet

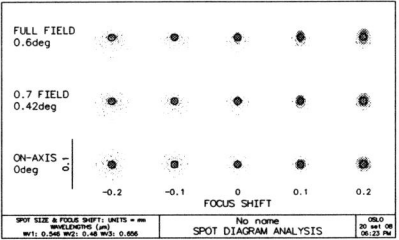

Fig. 7.9 Spot diagram for the Bk7-KZFS1-BaFN10 triplet

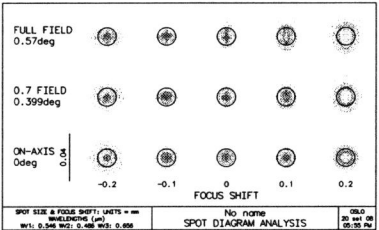

Fig. 7.10 Spot diagram for the Bk7-KZFS1-BaF4 triplet obtained using *Optisoft*

7.5 Examples

In this section we consider two examples of $f/10$ triplets of diameter $D = 100$ mm and speed $f/10$. The first is constructed from the glasses Bk7, KZFS1, and BaF4, and the second from Bk7, KZFS1, and BaFN10. Using

7.5. Examples   155

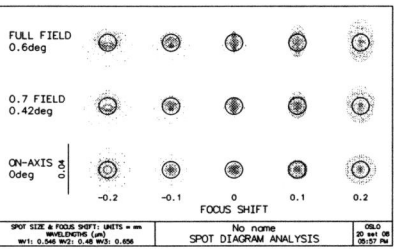

Fig. 7.11 Spot diagram for the Bk7-KZFS1-BaFN10 triplet obtained using *Optisoft*

the notebook *Triplet*, we obtain the following radii for the five surfaces of the lenses in each example triplet:

$$R_1 = 589.001, \quad R_2 = -210, \quad R_3 = -217.9,$$
$$R_4 = 211, \quad R_5 = -1185.6.$$

$$R_1 = 3951, \quad R_2 = -228.25, \quad R_3 = -220.3,$$
$$R_4 = 202.211, \quad R_5 = -605.55.$$

Figures 7.8 and 7.9 show the spot diagrams for these triplets. It is evident that they are not acceptable considering their higher-order aberrations. Figures 7.10 and 7.11 show the spot diagrams for the same triplets designed by *Optisoft*.

Chapter 8

Other Optical Combinations

8.1 Cassegrain Telescope with Spherical Surfaces

In this section we show that it is possible to realize an aplanatic Schmidt–Cassegrain telescope using only spherical mirrors. First, we suppose that all of the parameters for the Cassegrain combination are derived as shown in Chapter 4. That is, starting from the absolute value of the focal length f_1 of the primary mirror S_1, the required total focal length f_t of the whole combination, and the fixed distance em between S_1 and the focal plane (see Figure 8.1), we determine the focal length $f_2 = \gamma f_1$ of S_2 as well as the distance $-\beta f_1$ between the two mirrors using the relations

$$\alpha = \frac{em}{f_1}, \quad M = \frac{f_t}{f_1} \tag{8.1}$$

$$\beta = \frac{M-\alpha}{M+1}, \quad \gamma = \frac{2M(1+\alpha)}{1-M^2}. \tag{8.2}$$

Now we suppose that the aperture stop A with the Schmidt corrector is located at a distance $-\delta f_1$ in front of S_1 instead of being placed at S_1 (see Figure 8.1). The formulae for the spherical aberration and coma coefficients for this configuration can then be derived from the formulae we demonstrated in Chapter 6, provided that we put the conic constants of S_1 and S_2 equal to zero. Alternatively, we can resort to the notebook **Total-Aberrations**. In any case, we obtain the relations

$$D_1 = -\frac{1 + M^3 + \alpha - M^2(1+\alpha)}{32 f_1^3 M^3}, \tag{8.3}$$

$$D_2 = \frac{M + \alpha - M^3(\delta - 1) - \delta - \alpha\delta + M^2(\alpha(\delta-1)+\delta)}{8 f_1^2 M^3}. \tag{8.4}$$

Since the spherical aberration coefficient for the corrector C is $a_4(N-1)$, where a_4 is the aspherical coefficient of the second surface of C and N is

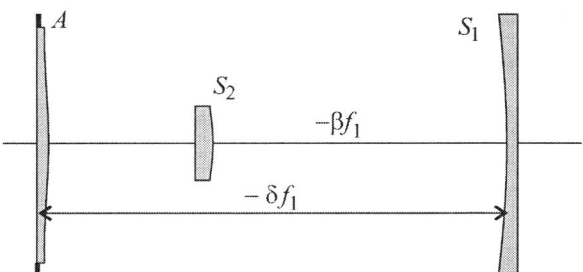

Fig. 8.1 Cassegrain combination with corrector

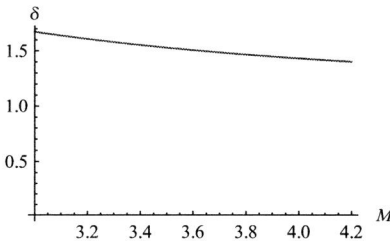

Fig. 8.2 Plot of δ versus M

the refractive index of C, the third-order spherical aberration is eliminated if we choose

$$a_4 = -\frac{D_1}{N-1}. \tag{8.5}$$

In order to eliminate the third-order coma, in view of (8.4), we can verify whether the root δ of the equation

$$M + \alpha - M^3(\delta - 1) - \delta - \alpha\delta + M^2(\alpha(\delta - 1) + \delta) = 0,$$

in other words, the value

$$\delta = \frac{M + M^3 + \alpha - M^2\alpha}{1 - M^2 + M^3 + \alpha - M^2\alpha}, \tag{8.6}$$

is acceptable or not. A plot of (8.6) for the typical value $\alpha = 0.33$ is shown in Figure 8.2.

For instance, for a mirror S_1 with a focal length $f_1 = 600$ mm, $\alpha = 0.33$ and $M = 4$, the third-order coma vanishes when $\delta = 1.4313$. Consequently, we must place the corrector at a distance of 858.78 from S_1. Figure 8.3 shows the corresponding spot diagram for this combination, and we can see that it is diffraction limited for a view angle of 0.24°.

8.2. The Flat-Field Baker–Schmidt Camera

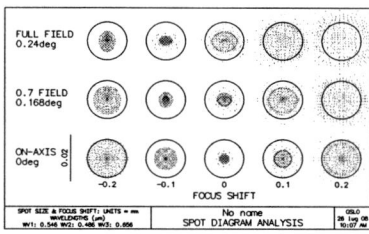

Fig. 8.3 Spot diagram for a Schmidt–Cassegrain combination

8.2 The Flat-Field Baker–Schmidt Camera

In this section we analyze some **flat-field Baker–Schmidt cameras**, as devised by Baker, Linfoot, and others (see [28, 29, 30, 31, 32, 33, 35]), which are free from spherical aberration, coma, and astigmatism. More precisely, we analyze cameras in which the primary mirror S_1 is spherical and the secondary mirror S_2 is conic.

Using the notation adopted in the above section (see Figure 8.1), we suppose that:

- The obstruction α due to S_2 is equal to 40% and so $\beta = 1 - \alpha = 0.6$. This obstruction is acceptable for photographic purposes.

- The back focal $bf > 0$, so that the focal surface π is located after the primary S_1.

This camera presents the following advantages over a Schmidt camera with the same speed:

- Its focal surface π is planar and external to the tube containing the whole combination

- It is shorter than the corresponding Schmidt camera.

We begin our analysis by considering the two mirrors along with an aperture stop a distance $-\delta f_1$ from S_1. Using the notebook **BakerSchmidt**, we then find that the total focal length f_t and the back focal bf are given by the following formulae when $\beta = 0.6$:

$$f_t = \frac{\gamma}{\gamma - 0.8} f_1, \qquad (8.7)$$

$$bf = \frac{0.48 - 0.2\gamma}{\gamma - 0.8} f_1. \qquad (8.8)$$

In what follows, we choose a value of γ that lies in the range

$$1 < \gamma < 2.4 \tag{8.9}$$

so that $bf > 0$.

With the help of the notebook **BakerSchmidt**, we then find that, for $\beta = 0.6$, the coefficient of curvature vanishes when the distance $-\delta f_1$ of the aperture stop from S_1 complies with the condition

$$-\delta f_1 = \frac{1.2 + \gamma}{\gamma - 0.8} f_1. \tag{8.10}$$

At this point, the geometry of the whole combination is determined by choosing a value of γ that satisfies condition (8.9). Moreover, the astigmatism coefficient D_3 vanishes for the corresponding value of δ given by (8.10), whereas the coefficients D_1 and D_2 (for the spherical aberration and coma, respectively) are given by the formulae

$$D_1 = \frac{1}{32 f_1^3 \gamma^3} \left(8 + 8K_2(-1+\beta)^4 + 8\beta^4 + 16\beta^3(-2+\gamma) - \right.$$
$$\left. 16\gamma + 8\gamma^2 - \gamma^3 + 8\beta^2(6 - 6\gamma + \gamma^2) - 16\beta(2 - 3\gamma + \gamma^2)\right), \tag{8.11}$$

$$D_2 = \frac{1}{8 f_1^2 \gamma^2 (-2 + 2\beta + \gamma)} \left(-8 - 8K_2(-1+\beta)^3 + 8\beta^3 + \right.$$
$$\left. 8\beta^2(-3+\gamma) + 8\gamma - 4\gamma^2 + \gamma^3 + 2\beta(12 - 8\gamma + \gamma^2)\right). \tag{8.12}$$

Ignoring the coma introduced by the corrector C and recalling that the coefficient of spherical aberration for C is given by $a_4(N-1)$, where a is the aspherical constant of C and N is its refractive index in green light, the third-order spherical aberration and coma are eliminated for a_4 and K_2 when the following system holds:

$$D_1 + a_4(N-1) = 0, \tag{8.13}$$
$$D_2 = 0, \tag{8.14}$$

where D_1 and D_2 are given by (8.11) and (8.12)). Solving the above system, we obtain

$$a_4 = -\frac{1}{32 f_1^3 (N-1)\gamma^3} \left[16\beta^4 - 2(-2+\gamma)^3 + \right.$$
$$\left. 8\beta^3(-8+3\gamma) + 2\beta^2(48 - 36\gamma + 5\gamma^2) + \right.$$
$$\left. \beta(-64 + 72\gamma - 22\gamma^2 + \gamma^3)\right], \tag{8.15}$$

$$K_2 = \frac{1}{8(-1+\beta)^3} \left[-8 + 8\beta^3 + 8\beta^2(-3+\gamma) + 8\gamma - \right.$$
$$\left. 4\gamma^2 + \gamma^3 - 2\beta(12 - 8\gamma + \gamma^2)\right]. \tag{8.16}$$

8.2. The Flat-Field Baker–Schmidt Camera

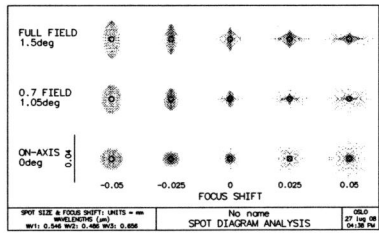

Fig. 8.4 Baker–Schmidt camera of speed $f/2$

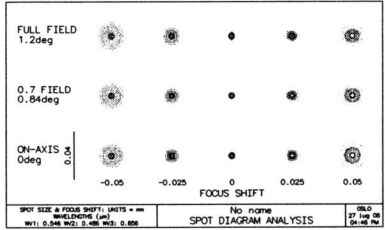

Fig. 8.5 Baker–Schmidt camera of speed $f/2$ obtained using ***Optisoft***

Note that all of the characteristics of the combination are functions of γ.

However, we cannot accept the design realized on the basis of the above formulae for the following reasons:

- The corrector introduces significant chromatism

- Combining the corrector and the mirrors leads to field curvature and significant astigmatism

- When the speed of the primary is high, higher-order aberrations degrade the final image.

We can overcome the first problem by assigning a suitable radius of curvature to the second surface of the corrector, as done in the program ***BakerSchmidt***, whereas the other problems require a deeper analysis.

Figure 8.4 shows the spot diagram obtained using the above formulae. The principal mirror has a diameter $D = 200$ mm and a speed of $f/2$. The camera has a total focal length of 628 mm. Figure 8.5 shows the spot diagram for the same camera designed by ***Optisoft***.

It is possible to realize a flat-field camera in which both mirrors are spherical (see [35]). The major disadvantage of this combination, beside the presence of a very small amount of astigmatism, is the large obstruction factor (about 55%). This combination is essentially based on the condition that the radii of curvature of the mirrors are equal. A full analysis of this configuration is provided in the notebook *BakerSchmidt*, together with some examples that exhibit different obstruction factors.

8.3 Cassegrain Telescope with the Corrector at the Prime Focus

In some of the preceding chapters, we analyzed optical devices in which the aberrations of a Cassegrain combination S were compensated for by introducing a suitable corrector C *before* S. When C is placed in this position, it is said to be a **full-aperture corrector**. It is evident that inserting C *after* S or inside S should reduce the size of C. In the rest of this chapter, we analyze two examples in which the corrector is located *after* the Cassegrain combination.

In Figure 8.6, a Schmidt corrector C is positioned after the two mirrors of the Cassegrain combination.

We show that this combination only works when both mirrors are *conic*. In fact, when one mirror is or both mirrors are spherical, the coma and astigmatism are significant and the aspherical constant a_4 of C is too high. The aberration formulae for this combination are very complex but *they are linear in the unknowns* a_4, K_1, *and* K_2, where K_1 and K_2 are the conic constants of S_1 and S_2, respectively. Since the corrector C does not modify the Gaussian behavior of the two mirrors, we analyze this combination starting from a given Cassegrain scheme. The notebook *NewCass* allows us to determine both the Cassegrain data and the unknowns a_4, K_1, and K_2 starting from the same input data for the Cassegrain combination.

We assume a principal mirror S_1 with a diameter $D = 200$ mm and an aperture ratio of $f/2$. If the requested total focal length is 2000 mm and the back focal $em = 200$ mm, then the data for the Cassegrain combination are (see *NewCass*):

$$\text{curvature radius of } S_1 = -800$$
$$\text{curvature radius of } S_2 = -250$$
$$\text{distance between } S_1 \text{and} S_2 = -300.$$

We now consider the combination S, which is obtained by placing the corrector C at a distance of 300 mm from S_2. In order to determine the

8.3. Cassegrain Telescope with the Corrector at the Prime Focus

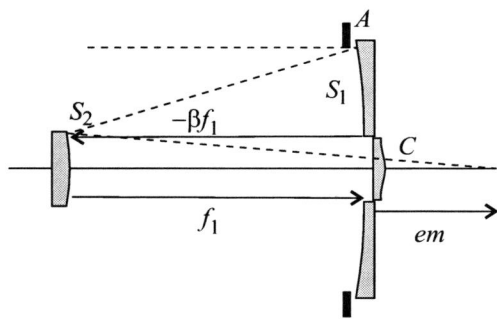

Fig. 8.6 New Schmidt–Cassegrain telescope

unknowns a_4, K_1, K_2, and the radius R_c of the second surface of C, which makes the corrector achromatic, we must equate the coefficients of spherical aberration D_1, coma D_2, and astigmatism D_3, as well as the difference Δ between the back distances for blue light and red light, to zero. To this end we use the notebook **TotalAberrations** (in **NewCass**) to verify that the condition $\Delta = 0$ is expressed by the equation

$$311.63 R_c + 0.0174 R_c^2 = 0, \tag{8.17}$$

which has a significant solution of

$$R_c = -17840.6. \tag{8.18}$$

Moreover, equating the corresponding expressions for D_1, D_2, and D_3 to zero, we obtain the linear system

$$D_1 = -3.47 + 485401 a_4 - 4.88 K_1 + 0.625 K_2 = 0, \tag{8.19}$$
$$D_2 = 7.369 + 2.62791 * 10^7 a_4 + 3 K_2 = 0, \tag{8.20}$$
$$D_3 = -0.000095 + 119555 a_4 + 0.00121 K_2 = 0, \tag{8.21}$$

which has the solution

$$a_4 = 2.81537 \times 10^{-8}, \quad K_1 = -1.05524, \quad K_2 = -2.70306. \tag{8.22}$$

Figure 8.7 shows the spot diagram for this combination produced on a focal surface with a radius of -200 mm.

164   Chapter 8. Other Optical Combinations

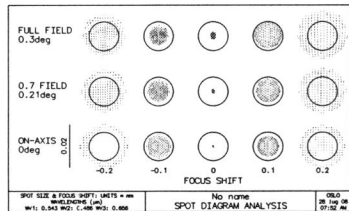

Fig. 8.7 Spot diagram for the Schmidt–Cassegrain telescope where the corrector is positioned after the mirrors

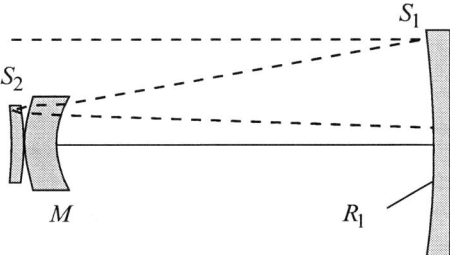

Fig. 8.8 Klevtsov telescope

8.4 The Klevtsov Telescope

In this section and the next two, we analyze another interesting combination where the corrector C is placed *after* the Cassegrain combination, which was devised by Klevtsov. The optical scheme for this combination is shown in Figure 8.8.

All of the surfaces are spherical. The secondary mirror S_2 is a Mangin mirror (i.e., only the second surface of S_2 is reflecting). The light crosses the meniscus and the secondary mirror S_2 (which we assume is constructed from BK7) twice. The disadvantages of this combination are as follows:

- The tube is open

- It requires very accurate collimation.

The analysis that we develop over the next few sections shows how useful ***Mathematica***® can be for solving the problems we encounter.

8.5. A First Analysis of the Meniscus

The formulae for the third-order aberrations of this combination are very complex, in terms of both the large number of surfaces involved and the finite thicknesses of its optical elements. As a consequence, we cannot hope to develop a theoretical analysis of this scheme starting from general formulae. Therefore, as we develop the design of the Klevtsov telescope, we assign the following parameters:

- The total focal length f_t

- The primary speed A

- The primary focal length f_1.

This leaves us with numerical equations where the radii of curvature of the surfaces involved are the unknowns.

To illustrate this approach, suppose that $f_t = 2200$ mm, $A = F/3$, and $f_1 = 600$ mm. Since we intend to use a Newton method to solve our equations, we need approximate values for our unknowns. To this end, we develop our design using a procedure based on a number of steps. Each of these steps yields approximate values for the unknowns that are closer to the final values. The process is encapsulated in the notebook **Klevtsov**, which simplifies the design of this combination. However, the final results obtained using this process could require improvement before they can be accepted.

8.5 A First Analysis of the Meniscus

In the first step of our process, we use the notebook **Cassegrain** to evaluate the Cassegrain combination on which the Klevtsov telescope is based. If we introduce the input data (see **Cassegrain**)

$$f_1 = 600, \quad f_t = 2200, \quad em = 240, \quad D = 200, \quad \theta = 0.26$$

for the primary focal length, the total focal length, the back focal, the diameter of the primary mirror, and the field angle, respectively, then we obtain the following output:

$$R_2 = -495, \quad \Delta = -420$$

for the radius R_2 of the secondary mirror S_2 and the distance Δ between S_1 and S_2. It is important to note that the meniscus reduces the back distance

166   Chapter 8. Other Optical Combinations

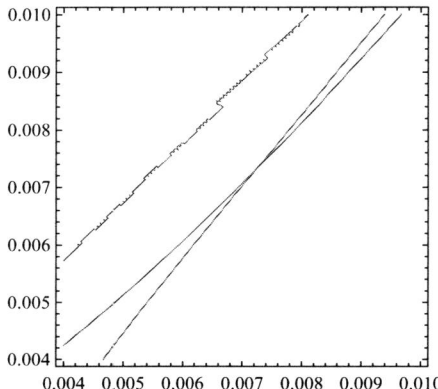

Fig. 8.9 The curves for which the spherical aberration and coma vanish

of the Cassegrain scheme. Therefore, we have fixed a back focal that is greater than the usual value of 200 mm.[1]

These data represent our starting data for designing the Klevtsov telescope. In the second step of the process, we analyze whether it is possible to eliminate the third-order spherical aberration and coma of the Cassegrain combination by choosing the curvatures c_3 and c_4 of the meniscus M appropriately. To this end, we consider a Klevtsov combination in which

- The primary has a radius $R_1 = 600$

- The Mangin mirror has *both* radii equal to -495 and a thickness of 3 mm

- The meniscus M has a thickness $\Delta_M = 30$ mm.

We use the notebook ***TotalAberrations*** to determine the very complex expressions for the third-order spherical aberration and coma in the above optical scheme. We do not try to write it out; instead we use the built-in function ***ContourPlot*** of ***Mathematica***® to get Figure 8.9, which shows the values of c_3 and c_4 for which the spherical aberration and coma vanish.

This figure allows us to determine approximate values for c_3 and c_4 to use as starting values in the built-in function ***FindRoot*** of ***Mathematica***® (see notebook ***Klevtsov***). In this way, we obtain

$$c_3 = 0.00723701, \quad c_4 = 0.00732246.$$

---

[1]The notebook ***Klevtsov*** uses a back focal em that is increased by the thickness of the meniscus.

8.6. Analysis of the Mangin Mirror

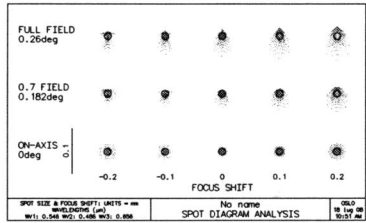

Fig. 8.10 Spot diagram for the Klevtsov telescope

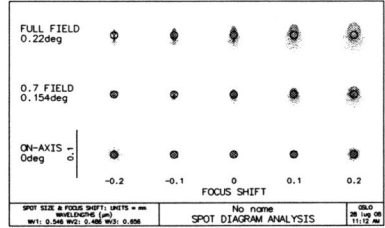

Fig. 8.11 Spot diagram for the Klevtsov telescope designed using *Optisoft*

The notebook *Klevtsov*) shows that spherical aberration and coma are absent but that the other data are not acceptable. In particular, we have a total focal length of 1830 mm. In the next section we show how to solve this problem.

8.6 Analysis of the Mangin Mirror

The third step of our design process involves varying the first curvature c_2 of the Mangin mirror as well as the curvatures of the meniscus in such a way that the third-order spherical aberration and coma are eliminated and a total focal length of 2200 mm is achieved. The above values of c_3, c_4, and $c_2 = -1/495$ are taken as starting values in *FindRoot*. In this way, we obtain

$$c_2 = -0.00163342, \quad c_3 = 0.00681001, \quad c_4 = 0.00683396.$$

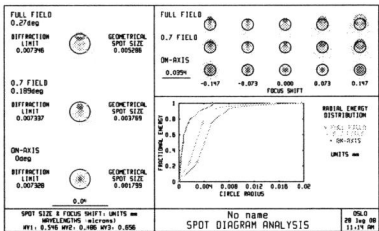

Fig. 8.12 Spot diagram for the Klevtsov telescope with a doublet, as designed using *Optisoft*

Analysis of the corresponding combination using the notebook ***TotalAberrations*** shows that third-order spherical aberration and coma are absent and that the total focal length is 2200 mm. However, there is consistent chromatism. In the fourth step we introduce as a new unknown the curvature of the primary mirror, c_1. Then, starting from the values of c_2, c_3, and c_4 that we have found, we again apply the built-in function ***FindRoot*** to the system obtained by eliminating the spherical aberration, coma, and axial chromatism for the given total focal length. This yields the final design, and the spot diagram for it is shown in Figure 8.10. We can see that the design must be optimized still further. Figure 8.11 shows the spot diagram for the same combination designed by *Optisoft*.

In any case, this combination is limited by its strong astigmatism and field curvature. However, by placing an achromatic doublet close to the end of the optical path, it is possible to achieve very good performance with this combination, as is illustrated by Figure 8.12, which refers to a Klevtsov telescope designed using *Optisoft*.

8.7 Buchroeder Camera

The ***Buchroeder camera*** employs a corrector C consisting of three thin lenses made from the same glass (for instance BK7), which is placed at the center of curvature of the primary mirror S. The radii of curvature of the six surfaces of the three lenses satisfy the condition

$$R_2 = \infty, \quad R_3 = -R_1, \quad R_4 = R_1, \quad R_5 = \infty, \quad R_6 = -R_1. \quad (8.23)$$

In other words, triplet C has a plane of symmetry (see Figure 8.13), and we assume that the aperture stop A lies within this plane.

8.7. Buchroeder Camera

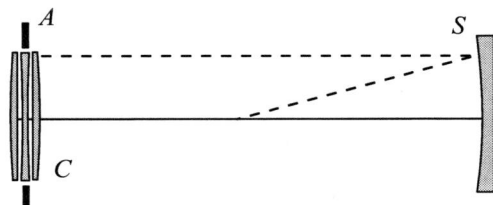

Fig. 8.13 Buchroeder camera

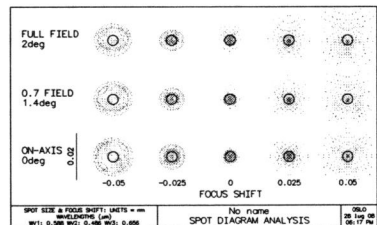

Fig. 8.14 $f/3$ Buchroeder camera

It is evident that, just like the Houghton corrector (see Chapter 6), the Buchroeder corrector is afocal *for any refractive index value*. Therefore, it does not disturb the Gaussian behavior of the mirror S. The system C–S has only one degree of freedom—the radius R_1. We can determine the value of R_1 by requiring that the spherical aberration vanishes. Using the notebook **Buchroeder**, we find that the corrector has no coma and astigmatism, whereas its coefficient of spherical aberration is given by

$$D_1 = \frac{c_1^3(N-1)^2(N+1)}{2N}, \tag{8.24}$$

where $c_1 = 1/R_1$ is the curvature of the surface of the first lens of C and N is its refractive index.

On the other hand, the mirror S (which has the aperture stop at its center of curvature) has no coma or astigmatism. Since its coefficient of spherical aberration is $-1/(32f^3)$, the absence of spherical aberration requires that

$$\frac{c_1^3(N-1)^2(N+1)}{2N} - 1/(32f^3) = 0. \tag{8.25}$$

The only real solution of the above equation is

$$c_1 = \frac{1}{2}\frac{N^{1/3}}{2^{1/3}f(N-1)^{2/3}(N+1)^{1/3}}. \tag{8.26}$$

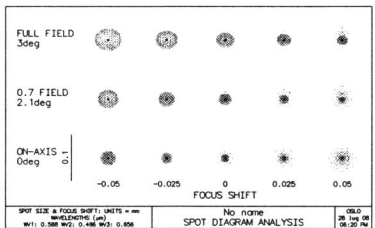

Fig. 8.15 $f/2$ Buchroeder camera

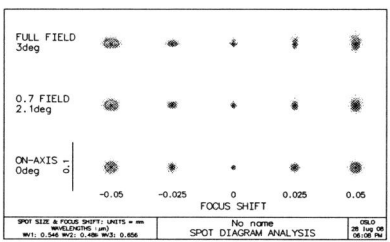

Fig. 8.16 $f/2$ Buchroeder camera designed using *Optisoft*

A camera based on this value of c_1 should be free of aberrations, except for the field curvature, provided that we can neglect the thickness of the lenses of C. The notebook *Buchroeder* allows us to design this kind of camera by modifying the value (8.26) in order to balance out the effect of the finite thickness of the lenses.

Figure 8.14 shows the spot diagram for a camera with an $f/3$ mirror, whereas Figure 8.15 shows a corresponding diagram for a camera with an $f/2$ mirror. Finally, Figure 8.16 shows the spot diagram for a camera with an $f/2$ mirror designed using *Optisoft*.

Chapter 9

Fermat's Principle and Wavefronts

9.1 Fermat's Principle

In Chapter 1 we introduced Fermat's principle in an elementary way, without any critical analysis of its consequences. In contrast, in this chapter, we will analyze Fermat's principle in depth, together with the related Lagrangian formalism and partial differential equations, in view of introducing the Hamiltonian formalism.

Let us consider the propagation of a monochromatic light wave in an isotropic inhomogeneous medium Σ. If the wavelength of the light is small compared with the dimensions of both the region in which the light propagates and the interposed obstacles (stops), then the propagation of the light can be described by **rays** (see Section 1.1). The refractive index $N(\mathbf{x})$ of the medium Σ is assumed to be a smooth function of the position vector $\mathbf{x} \in \Sigma$, except for the points belonging to a finite number of surfaces S_l, $l = 1, \ldots, k$, across which $N(\mathbf{x})$ exhibits finite discontinuities. We have already defined (see Section 1.1) the **optical path length** $OPL(\gamma_0)$ along a curve γ_0 using the following formula:

$$OPL(\gamma_0) = \int_{\gamma_0} N(\mathbf{x}) ds. \tag{9.1}$$

In Chapter 1, **Fermat's principle** was formulated in the following way:

The optical path length of a light ray from point $\overline{\mathbf{x}}$ to point $\overline{\mathbf{x}}'$ is the minimum among all potential optical path lengths corresponding to all potential paths between these points.

In this section we provide answers to the questions posed in Section 1.1.

In order to characterize the minimum among the potential optical path lengths, we first introduce a Cartesian coordinate frame (x_i), $i = 1, 2, 3$. Any path γ between $\overline{\mathbf{x}}$ and $\overline{\mathbf{x}}'$ is thus represented by three parametric equations $x_i = f_i(t)$, $i = 1, 2, 3$, $t \in [a, a']$, which give the coordinates

172   Chapter 9. Fermat's Principle and Wavefronts

$(f_i(t))$ of the points of γ as the parameter t varies. The points $\overline{\mathbf{x}} = (\overline{x}_i)$ and $\overline{\mathbf{x}}' = (\overline{x}'_i)$ will be the end points of γ provided that

$$f_i(a) = \overline{x}_i, \qquad f_i(a') = \overline{x}'_i. \tag{9.2}$$

In particular, the parametric equations for the *ray* γ_0 between $\overline{\mathbf{x}}$ and $\overline{\mathbf{x}}'$—in other words, for the *effective* light path from $\overline{\mathbf{x}}$ to $\overline{\mathbf{x}}'$—will be denoted by the functions $x_i(t)$, $i = 1, 2, 3$, with the conditions

$$x_i(a) = \overline{x}_i, \quad x_i(a') = \overline{x}'_i. \tag{9.3}$$

The next step is to search for the expression for the optical path length of any path γ between the points $\overline{\mathbf{x}}$ and $\overline{\mathbf{x}}'$, in order to determine the curve along with γ_0 has a minimum.

Analysis shows that the infinitesimal arc length ds of γ is given by

$$ds = \sqrt{\sum_{i=1}^{3} \dot{f}_i^2} dt, \tag{9.4}$$

where the notation $df_i/dt = \dot{f}_i$ has been used. Instead of (9.1), the following expression is now considered, which supplies the optical path length along any path γ between $\overline{\mathbf{x}}$ and $\overline{\mathbf{x}}'$:

$$OPL(\gamma) = \int_\gamma N(\mathbf{x}) \sqrt{\sum_{i=1}^{3} \dot{f}_i^2} dt. \tag{9.5}$$

Recalling known results, we derive the *necessary* conditions for the optical path length (9.5) to be minimized upon varying the curve $f_i(t)$. To this end, we consider the following arbitrary one-parameter family Γ of curves between the points $\overline{\mathbf{x}}$ and $\overline{\mathbf{x}}'$ that contain the effective ray γ_0:

$$x_i = f_i(t, \epsilon), \quad \forall t \in [a, a'], \qquad i = 1, 2, 3, \tag{9.6}$$

where
$$\begin{aligned} f_i(a, \epsilon) = \overline{x}_i, \qquad f_i(a', \epsilon) = \overline{x}'_i, \quad \forall \epsilon \in (-\delta, \delta) \\ f_i(t, 0) = x_i(t), \end{aligned} \tag{9.7}$$

and $(-\delta, \delta)$ denotes the neighborhood of the value $\epsilon = 0$ (see Figure 9.1).

The optical path length of a curve belonging to Γ depends on the curve chosen (i.e., on the value of ϵ) according to the following formula, which is obtained from (9.5):

$$OPL(\epsilon) = \int_a^{a'} N(\mathbf{x}) \sqrt{\sum_{i=1}^{3} \dot{f}_i^2(t, \epsilon)} dt. \tag{9.8}$$

9.1. Fermat's Principle

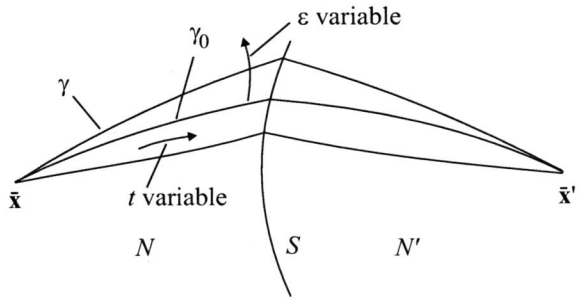

Fig. 9.1 The family Γ of optical paths

In other words, Fermat's principle can be equivalently stated as follows:
Determine the condition under which the function $OPL(\epsilon)$ has a minimum at $\epsilon = 0$ for any member of the family $(f_i(t, \epsilon))$ of curves, verifying (9.7).

A **necessary** condition to achieve this is that

$$OPL'(0) = 0, \qquad (9.9)$$

for any choice of Γ.

Consequently, we must

- Evaluate the first derivative of $OPL(\epsilon)$ at $\epsilon = 0$

- Determine the conditions under which it vanishes for any member of the family Γ

- Verify whether this condition characterizes the ray γ_0

- Check whether $OPL(0)$ is a minimum of $OPL(\epsilon)$.

A general situation is considered here. This will enable other important results to be derived in Chapter 11 without repeating similar calculations. More precisely, we evaluate the derivative at $\epsilon = 0$ of the following function:

$$I(\epsilon) = \int_{t_1(\epsilon)}^{t_2(\epsilon)} N(\mathbf{x}) \sqrt{\sum_{i=1}^{3} \dot{f}_i^2(t, \epsilon)} dt, \qquad (9.10)$$

where the family Δ of curves $f_i(t, \epsilon)$ comply with the following conditions instead of (9.6) and (9.7):

$$x_i = f_i(t, \epsilon), \quad \forall t \in [t_1(\epsilon), t_2(\epsilon)], \qquad i = 1, 2, 3, \qquad (9.11)$$

174 Chapter 9. Fermat's Principle and Wavefronts

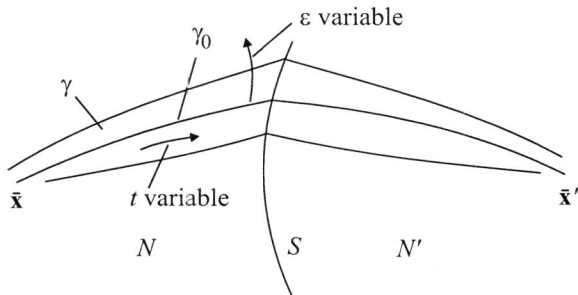

Fig. 9.2 A general family of optical paths

$$f_i(t_1(0), 0) = \overline{x}_i, \quad f_i(t_2(0), 0) = \overline{x}'_i, \quad \forall \epsilon \in (-\delta, \delta),$$
$$f_i(t, 0) = x_i(t), \tag{9.12}$$

where $(-\delta, \delta)$ is in the neighborhood of zero (see Figure 9.2).

In other words, the curves of Δ still contain the ray γ_0 for $\epsilon = 0$, but they have different initial and final points as ϵ varies. It is clear that the family Δ reduces to Γ, and so the function $I(\epsilon)$ reduces to $OPL(\epsilon)$, when

$$t_1(\epsilon) = a, \quad t_2(\epsilon) = a', \quad \forall \epsilon \in (-\delta, \delta). \tag{9.13}$$

Before determining the derivative $I'(\epsilon)$ and evaluating it at $\epsilon = 0$, it is important to note that:

1. The coordinates of the extreme points of the curves $\gamma \in \Delta$ are

$$f_i(t_1(\epsilon), \epsilon), \quad f_i(t_2(\epsilon), \epsilon), \tag{9.14}$$

so that, in going from the ray γ_0 (which corresponds to $\epsilon = 0$ in (9.12)) to the curve γ (corresponding to the value $d\epsilon$), the coordinates of the extreme points vary according to the relations

$$d\overline{x}_i = (\dot{x}_i \frac{dt_1}{d\epsilon} + \frac{\partial f_i}{\partial \epsilon})_{\epsilon=0} d\epsilon, \quad d\overline{x}'_i = (\dot{x}_i \frac{dt_2}{d\epsilon} + \frac{\partial f_i}{\partial \epsilon})_{\epsilon=0} d\epsilon. \tag{9.15}$$

2. The function $N(\mathbf{x})$ exhibits finite discontinuities across some of the surfaces S_i, $i = 1, \ldots, n$, included in the optical system. For the sake of simplicity, for the remainder of this section, it is assumed that there is a single discontinuous surface S. As a consequence, we hypothesize that the functions $f_i(t, \epsilon)$ may have discontinuous first derivatives; in other words, that the vector of the tangent to any path from Δ could abruptly vary in direction while crossing S.

9.1. Fermat's Principle

More precisely, if $F(\mathbf{x}) = 0$ denotes the equation of the surface S, the above discontinuities must be localized at the points of intersection of the curves of Δ with S. In order to determine the locus φ of these points, we first verify that the equation

$$F(f_i(t, \epsilon)) = 0, \tag{9.16}$$

implicitly defines a function $t = \psi(\epsilon)$ that gives the value of t corresponding to the point of intersection P between γ and S. In fact, Dini's condition

$$\frac{\partial}{\partial t} F(f_i(t, \epsilon)) \neq 0, \tag{9.17}$$

can be written

$$\frac{\partial F}{\partial x_i} \frac{\partial f_i}{\partial t} \neq 0, \tag{9.18}$$

or, equivalently,

$$\mathbf{n} \cdot \mathbf{t} \neq 0, \tag{9.19}$$

where $\mathbf{n} = \nabla F / |\nabla F|$ is the *internal* unit vector normal to S and $\mathbf{t}$ the unit vector tangent to γ at P. Therefore, if γ is not tangent to S at P, (9.16) implicitly defines a function $t = \psi(\epsilon)$, and the locus φ is a curve (see Figure 9.3) with parametric equations that yield

$$\varphi_i(\epsilon) = f_i(\psi(\epsilon), \epsilon), \tag{9.20}$$

where

$$F(\varphi_i(\epsilon)) = F(f_i(\psi(\epsilon), \epsilon)) = 0. \tag{9.21}$$

Due to the continuity of the curves of Δ while crossing S, $\varphi(\epsilon)$ is assumed to be a regular curve.

By differentiating (9.21) with respect to ϵ from both sides of S, and taking into account (9.20), the following two conditions are obtained:

$$\begin{aligned} n_i \left(\frac{\partial f_i}{\partial t} \frac{\partial \psi}{\partial \epsilon} + \frac{\partial f_i}{\partial \epsilon} \right)^- &\equiv n_i \xi_i^- = 0, \\ n_i \left(\frac{\partial f_i}{\partial t} \frac{\partial \psi}{\partial \epsilon} + \frac{\partial f_i}{\partial \epsilon} \right)^+ &\equiv n_i \xi_i^+ = 0, \end{aligned} \tag{9.22}$$

where the summation over the index i is intended and $()^+$, $()^-$ denote the limits of the quantities appearing between the parentheses when S is approached from the right-hand side and from the left-hand side, respectively.

Moreover, the regularity of $\varphi(\epsilon)$ implies the equality

$$\varphi_i'(\epsilon) = \left(\frac{\partial f_i}{\partial t} \frac{\partial \psi}{\partial \epsilon} + \frac{\partial f_i}{\partial \epsilon} \right)^- = \left(\frac{\partial f_i}{\partial t} \frac{\partial \psi}{\partial \epsilon} + \frac{\partial f_i}{\partial \epsilon} \right)^+. \tag{9.23}$$

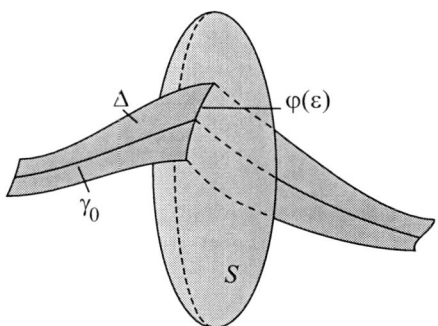

Fig. 9.3 The locus of the points of intersection between the optical paths and S

which can be written
$$\xi_i^- = \xi_i^+. \tag{9.24}$$
In what follows, the discontinuity of any quantity a across S will be denoted by
$$[[a]] = a^+ - a^-. \tag{9.25}$$
We are now able to evaluate the derivative $I'(0)$. First, we have
$$\begin{aligned} I(\epsilon) &= \int_{t_1(\epsilon)}^{t_2(\epsilon)} L(f_i(t,\epsilon), \dot{f}(t,\epsilon)) dt \\ &= \int_{t_1(\epsilon)}^{\psi(\epsilon)} L(f_i(t,\epsilon), \dot{f}_i(t,\epsilon)) dt \\ &+ \int_{\psi(\epsilon)}^{t_2(\epsilon)} L(f_i(t,\epsilon), \dot{f}_i(t,\epsilon)) dt, \end{aligned} \tag{9.26}$$
where
$$L = N(\mathbf{x}) \sqrt{\sum_{i=1}^{3}(\dot{f}_i)^2}. \tag{9.27}$$
By applying a well-known derivation formula for an integral, the above derivative becomes
$$\begin{aligned} \frac{dI}{d\epsilon}(0) = & L(\overline{x}_i', \dot{x}_i(a'))t_2'(0) - L(\overline{x}_i, \dot{x}_i(a))t_1'(0) \\ &+ \int_a^{\psi(0)} \left[\frac{\partial L}{\partial x_i}\frac{\partial f_i}{\partial \epsilon} + \frac{\partial L}{\partial \dot{x}_i}\frac{d}{dt}\left(\frac{\partial f_i}{\partial \epsilon}\right) \right] dt \\ &+ \int_{\psi(0)}^{a'} \left[\frac{\partial L}{\partial x_i}\frac{\partial f_i}{\partial \epsilon} + \frac{\partial L}{\partial \dot{x}_i}\frac{d}{dt}\left(\frac{\partial f_i}{\partial \epsilon}\right) \right] dt - [[L]]\psi'(0), \end{aligned}$$

9.1. Fermat's Principle

where all of the quantities that appear on the right-hand side are evaluated at $\epsilon = 0$. By integrating the second term in each integral in the above relation by parts, we get

$$\frac{dI}{d\epsilon}(0) = L(\overline{x}'_i, \dot{x}_i(a'))t'_2(0) - L(\overline{x}_i, \dot{x}_i(a))t'_1(0)$$
$$+ \int_a^{a'} \left[\frac{\partial L}{\partial x_i} - \frac{d}{dt}\frac{\partial L}{\partial \dot{x}_i}\right]\frac{\partial f_i}{\partial \epsilon} dt + \left[\frac{\partial L}{\partial \dot{x}_i}\frac{\partial f_i}{\partial \epsilon}\right]_a^{\psi(0)}$$
$$+ \left[\frac{\partial L}{\partial \dot{x}_i}\frac{\partial f_i}{\partial \epsilon}\right]_{\psi(0)}^{a'} - [[L]]\psi'(0), \qquad (9.28)$$

so that

$$\frac{dI}{d\epsilon}(0) = -\left(\frac{\partial L}{\partial \dot{x}_i}\dot{x}_i - L\right)_{a'} t'_2(0) + \left(\frac{\partial L}{\partial \dot{x}_i}\dot{x}_i - L\right)_a t'_1(0)$$
$$+ \int_a^{a'} \left(\frac{\partial L}{\partial x_i} - \frac{d}{dt}\frac{\partial L}{\partial \dot{x}_i}\right)\frac{\partial f_i}{\partial \epsilon} dt$$
$$+ \left(\frac{\partial L}{\partial \dot{x}_i}\left(\frac{\partial f_i}{\partial \epsilon} + \dot{x}_i t'_2\right)\right)_{a'} - \left(\frac{\partial L}{\partial \dot{x}_i}\left(\frac{\partial f_i}{\partial \epsilon} + \dot{x}_i t'_1\right)\right)_a$$
$$- \left[\left[\frac{\partial L}{\partial \dot{x}_i}\frac{\partial f_i}{\partial \epsilon}\right]\right]_{\psi(0)} - [[L]]\psi'(0). \qquad (9.29)$$

Finally, recalling (9.22) and (9.23), we get

$$\frac{dI}{d\epsilon}(0) = -\left(\frac{\partial L}{\partial \dot{x}_i}\dot{x}_i - L\right)_{a'} t'_2(0) + \left(\frac{\partial L}{\partial \dot{x}_i}\dot{x}_i - L\right)_a t'_1(0)$$
$$+ \int_a^{a'} \left(\frac{\partial L}{\partial x_i} - \frac{d}{dt}\frac{\partial L}{\partial \dot{x}_i}\right)\frac{\partial f_i}{\partial \epsilon} dt$$
$$+ \left(\frac{\partial L}{\partial \dot{x}_i}\right)_{a'}\frac{d\overline{x}'_i}{d\epsilon} - \left(\frac{\partial L}{\partial \dot{x}_i}\right)_a \frac{d\overline{x}_i}{d\epsilon}$$
$$- \left[\left[\frac{\partial L}{\partial \dot{x}_i}\left(\frac{\partial x_i}{\partial \epsilon} + \dot{x}_i\psi'(0)\right)\right]\right]_{\psi(0)}$$
$$- [[L - \frac{\partial L}{\partial \dot{x}_i}\dot{x}_i]]_{\psi(0)}\psi'(0). \qquad (9.30)$$

Since the function L is homogeneous to the first degree with respect to the variables $\dot{x}_i$, it follows that

$$L - \frac{\partial L}{\partial \dot{x}_i}\dot{x}_i = 0,$$

and the above formula becomes

$$\frac{dI}{d\epsilon}(0) = \int_a^{a'} \left(\frac{\partial L}{\partial x_i} - \frac{d}{dt}\frac{\partial L}{\partial \dot{x}_i}\right)\frac{\partial f_i}{\partial \epsilon} dt$$

$$+ \left(\frac{\partial L}{\partial \dot{x}_i}\right)_{a'} \frac{d\overline{x}'_i}{d\epsilon} - \left(\frac{\partial L}{\partial \dot{x}_i}\right)_{a} \frac{d\overline{x}_i}{d\epsilon}$$
$$- \left[\left[\frac{\partial L}{\partial \dot{x}_i}\left(\frac{\partial x_i}{\partial \epsilon} + \dot{x}_i \psi'(0)\right)\right]\right]_{\psi(0)}. \tag{9.31}$$

In order to apply Fermat's principle to (9.31), *we must assume that all of the curves have the same end points*. In this case, (9.31) reduces to the following equation:

$$\frac{dI}{d\epsilon}(0) = \int_a^{a'} \left[\frac{\partial L}{\partial x_i} - \frac{d}{dt}\frac{\partial L}{\partial \dot{x}_i}\right] \frac{\partial x_i}{\partial \epsilon} dt + \left[\left[\frac{\partial L}{\partial \dot{x}_i}\right]\right]_{\psi(0)} \xi_i, \tag{9.32}$$

where

$$\xi_i = \left(\frac{\partial x_i}{\partial \epsilon} + \dot{x}_i \psi'(0)\right)_{\psi(0)}$$

and (9.22) and (9.23) were used at $\epsilon = 0$.

Fermat's principle requires that the derivative of the optical path vanishes *for any choice of the quantities* $(\partial x_i/\partial \epsilon)_{\psi(0)}$ *and* ξ_i *such that*

$$\xi_i n_i = 0.$$

From the right-hand side of (9.32), it is easy to show that the above condition is equivalent to the following ones:

$$\frac{\partial L}{\partial x_i} - \frac{d}{dt}\frac{\partial L}{\partial \dot{x}_i} = 0, \tag{9.33}$$

$$\left[\left[\frac{\partial L}{\partial \dot{x}_i}\right]\right]_{\psi(0)} = \lambda n_i, \quad i = 1, 2, 3, \tag{9.34}$$

where λ is an unknown Lagrangian multiplier.

Equations 9.33 and 9.34 are, respectively, the **Euler–Lagrange equations** and the **Weierstrass–Erdman jump conditions** for the functional

$$\int_a^{a'} L\,dt = \int_a^{a'} N\sqrt{\sum_{i=1}^{3}(\dot{x}_i)^2}\,dt. \tag{9.35}$$

In order to convert (9.33) and (9.34) into a more expressive form, it is convenient to introduce the unit vector tangent to the ray γ_0, which propagates from the point $\overline{x}$ to $\overline{x}'$:

$$\mathbf{t} = \frac{\dot{x}_i}{\sqrt{\sum_{i=1}^{3}(\dot{x}_i)^2}}, \tag{9.36}$$

9.1. Fermat's Principle

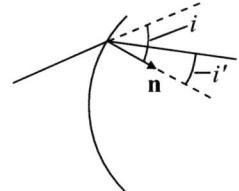

Fig. 9.4 The refraction law

and to recall the relation

$$\frac{ds}{dt} = \sqrt{\sum_{i=1}^{3}(\dot{x}_i)^2}. \tag{9.37}$$

It is then straightforward to verify that (9.33) yields

$$\frac{d}{ds}(N\mathbf{t}) = \nabla \mathbf{N}. \tag{9.38}$$

In particular, this equation implies that *light propagates in straight lines in any region with a constant refractive index N*.

Using this notation, condition (9.34) can be written as follows:[1]

$$N'\mathbf{t}' - N\mathbf{t} = \lambda\mathbf{n}, \tag{9.39}$$

so that the incident ray, the refracted ray, and the unit vector **n** that is normal to S and oriented in the direction of propagation of the light are *coplanar* at the incidence point. Moreover, vector multiplication of (9.39) with **n** leads to the relation

$$N'\mathbf{t}' \times \mathbf{n} = N\mathbf{t} \times \mathbf{n}, \tag{9.40}$$

which in turn yields

$$N' \sin i' = N \sin i, \tag{9.41}$$

where i and i' are the angles of incidence and refraction, respectively (i.e., the angles that **t** and **t'** form with **n**; see Figure 9.4). The property that **t**, **t'**, and **n** are coplanar as well as relation (9.41) represent the **refraction law** of geometrical optics.

It is now possible to evaluate the Lagrangian multiplier λ. In fact, the scalar product between **n** and (9.39) supplies the relation

$$\lambda = N' \cos i' - N \cos i = N'\sqrt{1 - \sin^2 i'} - N \cos i,$$

---
[1] See Chapter 1 and Exercise 1 at the end of Chapter 1.

which, in view of (9.41), yields

$$\lambda = \sqrt{N'^2 - N^2 \sin^2 i} - N \cos i. \tag{9.42}$$

We have now shown that Fermat's principle includes all of the properties of light rays. In particular, (9.41) and the fact that $N' = -N$ applies to reflection result in the **reflection law**: *the incident and reflected rays are coplanar with* **n** *and subtend the same angle with* **n**.

Bearing in mind what we have said so far, Fermat's principle can be more precisely stated as follows:

The light ray γ_0 between two points **x** *and* **x**$'$ *is an **extremum** of the optical path function $OPL(\epsilon)$, although not necessarily a minimum.*

9.2 The Boundary Value Problem

It is worth noting that the determination of the light ray γ_0 between two points **x** and **x**$'$ by Fermat's principle has now been reduced to a search for the solution $(x_i(t))$, $i = 1, 2, 3$, to the differential equation (9.33) that satisfies the boundary conditions of (9.3) as well as the jump condition of (9.34); in other words

$$\frac{\partial L}{\partial x_i} - \frac{d}{dt}\frac{\partial L}{\partial \dot{x}_i} = 0,$$
$$x_i(a) = x_a^i, \quad x_i(b) = x_b^i, \tag{9.43}$$
$$\left[\left[\frac{\partial L}{\partial \dot{x}_i}\right]\right]_{\psi(0)} = \lambda n_i, \quad i = 1, 2, 3.$$

It has already been noted that this problem is very different from Cauchy's initial value problem. There are well-known existence and uniqueness theorems that apply under very general conditions for the latter. However, there is no known general theorem of this kind for the boundary value problem of (9.43).

Using examples, it is possible to verify that the ray γ_0 can be an (absolute or relative) minimum, a maximum, or neither. For instance, during reflection at a plane mirror, the ray **xx*****x**$'$ (where **x*** is the point of incidence) is a relative minimum, whereas the direct ray from **x** to **x**$'$ is an absolute minimum of this optical path.

The exercise at the end of Chapter 1 on conic mirrors shows that the boundary value problem of (9.43) can have an infinite number of solutions.

9.2. The Boundary Value Problem

In order to gain a deeper insight into this problem, we recall that a curve of $\Re^3$ is defined as a set of three functions $t \in [a, b] \to (x_1(t), x_2(t), x_3(t))$. Consequently, a change in the parameter $t = t(\tau)$ generates a *new* curve. In other words, a curve is a mapping of $t \to \Re^3$, not a locus of points of $\Re^3$. On the other hand, (9.5) is independent of the choice of parameter for t since Fermat's principle defines a ray as locus of points. Correspondingly, the boundary value problem (9.43) must define the solution *up to a change in the parameter*, so the equations of (9.43) cannot be *independent*. In fact, the identity (summation over i)

$$L - \dot{x}_i \frac{\partial L}{\partial \dot{x}_i} = 0$$

implies that

$$\frac{\partial L}{\partial x_h} = \dot{x}_i \frac{\partial^2 L}{\partial \dot{x}_i x_h}, \quad \frac{\partial L}{\partial \dot{x}_h} = \frac{\partial L}{\partial \dot{x}_h} + \dot{x}_i \frac{\partial^2 L}{\partial \dot{x}_i \dot{x}_h}, \tag{9.44}$$

and so

$$\dot{x}_i \frac{\partial^2 L}{\partial \dot{x}_i \dot{x}_h} = 0. \tag{9.45}$$

Due to (9.44) and (9.45), we can easily verify that

$$\dot{x}_i \left(\frac{\partial L}{\partial x_i} - \frac{d}{dt} \frac{\partial L}{\partial \dot{x}_i} \right) = 0$$

and so one of Lagrange's equations is verified as a consequence of the others.

In conclusion, one of the three Lagrange equations in (9.43) can be eliminated. In fact, since the solutions to these equations are independent of the parameter, we can assume that it is possible to introduce a coordinate frame in which all of the rays of interest can be represented by the following parametric equations:

$$x_\alpha = x_\alpha(x_3), \quad \alpha = 1, 2. \tag{9.46}$$

According to this hypothesis, we have

$$L = N(x_i)\sqrt{1 + \dot{x}_1^2 + \dot{x}_2^2}, \tag{9.47}$$

where the dot denotes *differentiation with respect to* x_3, and the independent equations (9.43) are

$$\frac{\partial L}{\partial x_\alpha} - \frac{d}{dx_3} \frac{\partial L}{\partial \dot{x}_\alpha} = 0,$$

$$x_\alpha(a) = x_a^\alpha, \quad x_\alpha(b) = x_b^\alpha, \tag{9.48}$$

$$\left[\left[\frac{\partial L}{\partial \dot{x}_\alpha}\right]\right]_{\psi(0)} = \lambda n_\alpha, \quad \alpha = 1, 2,$$

where λ is given by (9.42).

Let x_3^* be the value of the parameter corresponding to the point of intersection of the ray with the surface S, and let $x_\alpha^1(x_3, A_\alpha, B_\alpha), x_\alpha^2(x_3, C_\alpha, D_\alpha)$ be the general integrals of (9.48) in the intervals (a, x_3^*) and (x_3^*, b), respectively. In order to find the solution to boundary value problem (9.48), the eight constants $A_\alpha, B_\alpha, C_\alpha, D_\alpha$ must be determined. To achieve this, the six equations of (9.48) together with the two continuity conditions

$$x_\alpha^1(x_3^*, A_\alpha, B_\alpha) = x_\alpha^2(x_3^*, C_\alpha, D_\alpha), \tag{9.49}$$

are (at least in principle) sufficient.

Remark Let S be a mechanical system with holonomic, fixed and smooth constraints, upon which conservative forces deriving from the potential energy U act. The system S is assumed to have n degrees of freedom, and we denote its configuration space by V_n. This means that the configurations of S are in a one-to-one correspondence with the points of V_n. Then, if the Lagrangian coordinates $\mathbf{q} = (q_1, \ldots, q_n)$ are introduced in V_n, the motion of S will be geometrically described by a curve $\mathbf{q}(t) = (q_1(t), \ldots, q_n(t))$ of V_n. These curves are defined by another variational principle. In order to attain its formulation, we introduce the Lagrangian kinetic energy

$$T = \frac{1}{2} a_{hk}(\mathbf{q}) \dot{q}_h \dot{q}_k \tag{9.50}$$

and consider the class of motions that have a pre-fixed but arbitrary value E of energy. Moreover, we consider the Riemannian manifold obtained by equipping V_n with the metrics

$$d\sigma^2 = 2(E - U(\mathbf{q})) a_{hk}(\mathbf{q}) dq_h dq_\kappa. \tag{9.51}$$

Maupertuis' principle can be formulated as follows:

Let S be a mechanical system with holonomic, fixed and smooth constraints that is subjected to forces deriving from the potential energy U. The curves of V_n, representing motions of S with energy E, between two configurations $\mathbf{q}_1$ and $\mathbf{q}_2$, are geodesics of the Riemannian metrics (9.51); i.e., extrema of the functional

$$\int_{\mathbf{q}_1}^{\mathbf{q}_2} d\sigma. \tag{9.52}$$

Moreover, the point P (representing S in V_n) moves along these curves according to the following law:

$$\frac{d\sigma}{dt} = 2(E - U). \tag{9.53}$$

9.3. Rotational Symmetry and Lagrange's Invariant

In particular, if S is a material point P with mass m and orthogonal Cartesian coordinates (x_i) are introduced, we have

$$T = \frac{1}{2} \sum_{i=1}^{3} (\dot{x}_i)^2,$$

and so (9.51) and (9.52), respectively, yield

$$d\sigma^2 = 2(E - U) \sum_{i=1}^{3} dx_i^2, \qquad \int_{\mathbf{x}_1}^{\mathbf{x}_2} \sqrt{2(E - U) \sum_{i=1}^{3} dx_i^2}, \qquad (9.54)$$

where $\mathbf{x}_1, \mathbf{x}_2$ are the initial and final positions of P.

A comparison between (9.35) and (9.54) shows that, for fixed initial and final configurations, the material point P, when subjected to forces deriving from the potential energy U, follows the same trajectory as a light ray between the same points provided that

$$\sqrt{2(E - U)} = N.$$

It is well known that this perfect correspondence between mechanics and optics is one of the foundations of quantum mechanics.

9.3 Rotational Symmetry and Lagrange's Invariant

Let $\mathbb{S}$ be an **axially symmetric optical system** (i.e., a system with a rotational symmetry around an axis a), and let us choose the frame of reference $Ox_1x_2x_3$ in such a way that $x_3 \equiv a$. If polar coordinates (ρ, θ), where

$$\begin{aligned} x_1 &= \rho \cos \theta, & \rho &= \sqrt{x_1^2 + x_2^2}, \\ x_2 &= \rho \sin \theta, & \theta &= \arctan \frac{x_2}{x_1}, \end{aligned} \qquad (9.55)$$

are introduced in the plane Ox_1x_2, then, due to the symmetry, the Lagrangian function (9.47) of $\mathbb{S}$ yields

$$L = N(\rho, x_3)\sqrt{1 + \dot{\rho}^2 + \rho^2 \dot{\theta}^2}. \qquad (9.56)$$

Since L does not depend on θ, the quantity

$$I \equiv \frac{\partial L}{\partial \dot{\theta}} = N \frac{\rho^2 \dot{\theta}}{\sqrt{1 + \dot{\rho}^2 + \rho^2 \dot{\theta}^2}} \qquad (9.57)$$

is independent of x_3. Again expressing this quantity in terms of Cartesian coordinates, we obtain

$$I = N\frac{\dot{x}_2 x_1 - \dot{x}_1 x_2}{\sqrt{1 + \dot{x}_1^2 + \dot{x}_2^2}}, \qquad (9.58)$$

which, taking into account (9.36) and introducing the notation

$$p_i = N\frac{\dot{x}_i}{\sqrt{1 + \dot{x}_1^2 + \dot{x}_2^2}}, \qquad (9.59)$$

becomes

$$I = p_2 x_1 - p_1 x_2 = \text{const}. \qquad (9.60)$$

In other words, the quantity I is constant along a ray *in any region in which the refractive index is constant*.

We now verify that I is an *optical invariant*; in other words, that it is constant along any ray, regardless of the reflections and refractions it undergoes. In fact, for any surface S that is discontinuous in the refractive index N, in view of (9.59), the discontinuity condition of (9.48)$_3$ can be written as follows:

$$[[p_\alpha]] = \lambda n_i. \qquad (9.61)$$

Consequently, since $x_i^+ = x_i^-$, when the ray meets S we have

$$p_2^+ x_1^+ - p_1^+ x_2^+ = p_2^- x_1^- - p_1^- x_2^- + \lambda(x_1^- n_2 - x_2^- n_1), \qquad (9.62)$$

because of the continuity of the ray. On the other hand, if S exhibits rotational symmetry around the axis x_3, we have $x_3 = f(x_1^2 + x_2^2) \equiv f(\rho)$ and so the components n_1 and n_2 of its unit normal vector become

$$\begin{aligned} n_1 &= \frac{f_{,x_1}}{\sqrt{1 + (f_{,x_1})^2 + (f_{,x_2})^2}} = \frac{2f'x_1}{\sqrt{1 + (f_{,x_1})^2 + (f_{,x_2})^2}}, \\ n_2 &= \frac{f_{,x_2}}{\sqrt{1 + (f_{,x_1})^2 + (f_{,x_2})^2}} = \frac{2f'x_2}{\sqrt{1 + (f_{,x_1})^2 + (f_{,x_2})^2}}, \end{aligned} \qquad (9.63)$$

where $f_{,x_i} = \partial f/\partial x_i$ and $f' = df/d\rho$. Introducing these expressions into (9.62), we can easily derive

$$I^+ = p_2^+ x_1^+ - p_1^+ x_2^+ = p_2^- x_1^- - p_1^- x_2^- = I^-, \qquad (9.64)$$

and so the complete invariance of I along a ray is proven. The quantity I is called **Lagrange's optical invariant**.

In the next chapter this invariant will be derived in the Hamiltonian formalism, together with other aspects of symmetric optical systems.

9.4 Wavefronts and Fermat's Principle

Geometrical optics requires only the concepts of rays or wavefronts, which are surfaces that are orthogonal to rays that move with normal velocity c/N. In this section, we prove the equivalence of these two approaches without presenting the physical background for propagation.

First, we note that equations (9.38), i.e.,

$$\frac{d}{ds}\left(N\frac{dx_i}{ds}\right) = \frac{\partial N}{\partial x_i},$$

are equivalent to the first-order differential system

$$\frac{dx_i}{ds} = \frac{p_i}{N}, \quad (9.65)$$

$$\frac{dp_i}{ds} = \frac{\partial N}{\partial x_i}. \quad (9.66)$$

Let us suppose that the propagation of light is described by wavefronts $g(x_i, t) = 0$ with a velocity of c/N:

$$-\frac{\partial g}{\partial t}\frac{1}{|\nabla g|} = \frac{c}{N}. \quad (9.67)$$

If the family of wavefronts is assumed to take the form $\psi(x_i, t) - ct = 0$, the above equation can be written as follows:

$$|\nabla \psi| = N$$

or equivalently as

$$\sum_{i=1}^{3}\left(\frac{\partial \psi}{\partial x_i}\right)^2 = N^2. \quad (9.68)$$

Equation 9.68 is called the **eikonal equation**, and it is a nonlinear partial differential equation in the unknown $\psi(x_i)$.

We are interested in the following Cauchy problem for (9.68):

Determine a solution $\psi(x_i)$ of (9.68) that assumes a constant value ψ_0 on an assigned surface Σ_0

$$x_{0i} = x_{0i}(\xi_\alpha), \quad \alpha = 1, 2,$$

where ξ_α are surface parameters.

The standard way of solving such a problem involves resorting to the *characteristic system* of (9.68). In fact, the solutions to this system, which are called the *characteristic curves* of the eikonal equation, allow us to derive the solution to Cauchy's problem.

The characteristic system yields

$$\frac{dx_i}{d\tau} = 2p_i, \qquad (9.69)$$

$$\frac{dp_i}{d\tau} = 2N\frac{\partial N}{\partial x_i}, \qquad (9.70)$$

$$\frac{d\psi}{d\tau} = 2N^2, \qquad (9.71)$$

where

$$p_i = \frac{\partial \psi}{\partial x_i}. \qquad (9.72)$$

Before we go any further, we note that an important conclusion can be derived from these preliminary considerations. Let us introduce the parameter s via the relation $ds = 2N d\tau$. If $(x_i(\tau), p_i(\tau))$ is a solution of (9.69) and (9.70), the curvilinear abscissa along the curve $(x_i(\tau))$ is given by

$$ds = \sqrt{\sum_{i=1}^{3} \left(\frac{dx_i}{d\tau}\right)^2} \, d\tau = 2\sqrt{\sum_{i=1}^{3} p_i^2} \, d\tau,$$

which means that—considering (9.72) and (9.68)—we have

$$ds = 2N d\tau.$$

In other words, s coincides with the curvilinear abscissa along the curve $(x_i(\tau))$.

In terms of the parameter s, system (9.69)–(9.71) becomes:

$$\frac{dx_i}{ds} = \frac{p_i}{N}, \qquad (9.73)$$

$$\frac{dp_i}{ds} = \frac{\partial N}{\partial x_i}, \qquad (9.74)$$

$$\frac{d\psi}{ds} = N. \qquad (9.75)$$

Equations (9.73)–(9.75) are identical to (9.65)–(9.66). Taking into account (9.72), we can therefore conclude that:

- The curves $x_i(s)$ that solve the eikonal equation coincide with the corresponding rays

- The rays are orthogonal to the surfaces $\psi = $ const.

9.4. Wavefronts and Fermat's Principle

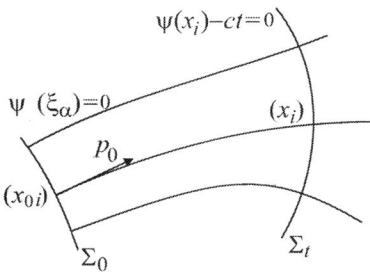

Fig. 9.5 Rays from Σ_0

In order to solve system (9.69)–(9.71), we need initial data. However, the points $x_{0i} = x_{0i}(\xi_\alpha)$ on the given surface Σ_0 supply the initial data for the unknowns x_i, but not for p_i. The initial data $p_{0i} = p_{0i}(\xi_\alpha)$ to use for this other set of unknowns can be obtained from the following conditions:

$$\sum_{i=1}^{3}(p_{0i}(\xi_\alpha))^2 = N^2(x_{0i}(\xi_\alpha)), \qquad (9.76)$$

$$\sum_{i=1}^{3}\frac{\partial x_{0i}(\xi_\alpha)}{\partial \xi_\alpha}p_{0i}(\xi_\alpha) = 0. \qquad (9.77)$$

The first condition, when (9.72) is taken into account, guarantees that the initial values of p_i satisfy the eikonal equation; the second one ensures that the vector (p_{0i}) is orthogonal to Σ_0 at any point. The two equations (9.76) and (9.77) can be solved with respect to the variables p_{0i}. The Jacobian matrix on Σ_0 is

$$J = \begin{pmatrix} 2p_{0i}(\xi(\alpha)) \\ \dfrac{\partial x_{0i}(\xi_\alpha)}{\partial \xi_\alpha} \end{pmatrix}.$$

However, at any point on Σ_0, the vector (p_{0i}) is orthogonal to the vectors $\frac{\partial x_{0i}(\xi_\alpha)}{\partial \xi_\alpha}$, which are tangent to Σ_0. Therefore, we have

$$J \neq 0$$

and (9.76) and (9.77) can be solved with respect to (p_{0i}) (see Figure 9.5).
Let

$$x_i(x_{0i}(\xi_\alpha), s) = \hat{x}_i(\xi_\alpha, s), \qquad (9.78)$$
$$p_i(p_{0i}(\xi_\alpha), s) = \hat{p}_i(\xi_\alpha, s) \qquad (9.79)$$

denote the solutions of (9.76) and (9.77), starting from any point on the surface Σ_0. Substituting this into (9.75), we finally obtain

$$\psi(\xi_\alpha, s) = \psi_0 + \int_0^s N(\hat{x}_i(\xi_\alpha, s))ds. \qquad (9.80)$$

Upon solving (9.78) with respect to the variables ξ_α, s, (9.80) becomes

$$\psi(x_i) = \psi_0 + \int_{x_0}^x N(\hat{x}_i(x_i))ds, \qquad (9.81)$$

where the integration is performed along a ray. From (9.81), we find that:

The difference between the values at two points of a solution to the eikonal equation is equal to the optical path length along the ray between these points.

The eikonal equation shows that the first derivatives of ψ are discontinuous across a surface S that separates an optical medium from another one. This means that S is a singular surface in ψ. Applying Maxwell's theorem, we can write

$$\left[\left[\frac{\partial \psi}{\partial x_i}\right]\right] = \lambda n_i. \qquad (9.82)$$

Taking into account (9.72), this condition becomes

$$[[p_i]] = \lambda n_i, \qquad (9.83)$$

and the refraction law is obtained.

Conversely, if the rays are known, it is sufficient to introduce the function (9.81) in order to obtain a family $\psi(x_i, t) - ct = 0$ of moving surfaces orthogonal to the rays of velocity c/N. Moreover, condition (9.82) can be derived from the refraction law. Finally, as shown by (9.81), the $\psi(x_i)$ satisfy the eikonal equation.

9.5 Huygens' Principle

Let Σ_0 be a wavefront at the initial time with an equation of $\psi(\xi_\alpha) = 0$. In the above section we described how to determine the wavefront Σ_t at the instant t, as represented by the equation $\psi(x_i) - ct = 0$. It is sufficient to integrate the equations (9.73) and (9.74) of the characteristic system by utilizing initial data consisting of the points $x_{0i}(\xi_\alpha)$ of Σ_0 and the vectors $p_{0i}(\xi_\alpha)$ orthogonal to Σ_0 and verifying condition (9.76). The function $\psi(x_i)

9.5. Huygens' Principle

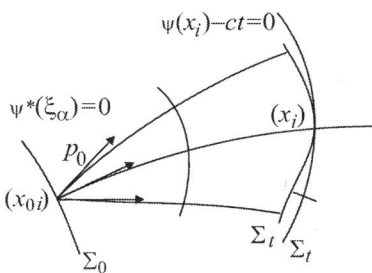

Fig. 9.6 Huygens' principle

is obtained using (9.81); the integration is performed along the rays from Σ_0 that have initial directions of $p_{0i}(\xi_\alpha)$ (see Figure 9.5).

In this section, a different way to obtain the wavefront Σ_t is described. To this end, an arbitrary but *fixed* point $(x_{0i}) \in \Sigma_0$ is chosen and the set of *all* vectors (p_{0i}) at (x_{0i}) is considered, with the condition (9.76) verified (see Figure 9.6).

By integrating the equations (9.73) and (9.74) of the characteristic system, and utilizing the initial data x_{0i} and p_{0i}, where (p_{0i}) is arbitrary to within the condition (9.76), the set Γ of the rays from (x_{0i}) can be obtained. Let us represent the wavefront Σ_t^* at the instant t, corresponding to propagation along these rays, by the equation $\psi^*(\xi_\alpha, x_i) - ct = 0$, where $\psi^*(\xi_\alpha, x_i)$ is always given by (9.81), but in this case the integration is performed *along the rays of* Γ. As a consequence, ψ^* verifies the eikonal equation.

Huygens' principle states that:

The wavefront Σ_t is the envelope of all of the surfaces Σ_t^* upon varying $(x_{0i}) \in \Sigma_0$.

To prove this statement, it is sufficient to recall that the envelope of the wavefronts $\psi^*(\xi_\alpha, x_i) - ct$ upon varying (x_{0i}) (i.e., the parameters ξ_α) can be obtained by eliminating these parameters in the system

$$\psi^*(\xi_\alpha, x_i) - ct = 0, \tag{9.84}$$

$$\frac{\partial \psi^*}{\partial \xi_\alpha}(\xi_\alpha, x_i) = 0. \tag{9.85}$$

More precisely, if we assume that (9.85) can be solved with respect to the parameters ξ_α,

$$\xi_\alpha = \xi_\alpha(x_i),$$

substitution of these equations into (9.84) yields

$$\psi^*(\xi_\alpha, x_i) - ct \equiv \psi(x_i) - ct = 0. \tag{9.86}$$

To prove that the function $\psi(x_i)$ in (9.86) is a solution to the eikonal equation, we note that we can derive the following from (9.85):

$$\frac{\partial \psi}{\partial x_i} = \frac{\partial \psi^*}{\partial \xi_\alpha} \frac{\partial \xi_\alpha}{\partial x_i} + \frac{\partial \psi^*}{\partial x_i} = \frac{\partial \psi^*}{\partial x_i}.$$

This result states that, at the point x_i, the surface $\psi(x_i) - ct = 0$ has the same tangent plane as the wave $\psi^*(\xi_\alpha, x_i) - ct = 0$, which originates from $x_{0i}(\xi_\alpha)$. However, $\psi*$ satisfies the eikonal equation, and this is true for $\psi(x_i)$.

Chapter 10

Hamiltonian Optics

10.1 Hamilton's Equations in Geometrical Optics

In Chapter 9, we formulated the fundamental laws of geometrical optics using the Lagrangian formalism. It is well known that any theory expressed in Lagrangian terms can also be (equivalently) formulated via the Hamiltonian formalism, with significant advantages. This chapter describes this new formalism together with its consequences.

First, we again suppose that it is possible to introduce coordinate frame in which all of the rays of interest can be represented by the parametric equations of (9.46). In this hypothesis, the two independent Lagrangian equations are written in the form of (9.48), where the Lagrangian $L(x_\alpha, \dot{x}_\alpha, x_3)$, $\alpha = 1, 2$, is given by (9.47) and *the dot expresses the derivation with respect to the variable x_3*.

As usual, the kinetic momenta are defined by the relations

$$p_\alpha = \frac{\partial L}{\partial \dot{x}_\alpha} = N \frac{\dot{x}_\alpha}{\sqrt{1 + \dot{x}_1^2 + \dot{x}_2^2}}. \tag{10.1}$$

Before introducing the Hamiltonian function, we must express the quantities $\dot{x}_\alpha$ in terms of p_α. Since

$$p_3^2 = N^2 - p_1^2 - p_2^2 = \frac{N^2}{1 + \dot{x}_1^2 + \dot{x}_2^2},$$

we have

$$\frac{N}{\sqrt{1 + \dot{x}_1^2 + \dot{x}_2^2}} = \sqrt{N^2 - p_1^2 - p_2^2}. \tag{10.2}$$

Therefore, (10.1) lead us to the relations

$$\dot{x}_\alpha = \frac{p_\alpha}{\sqrt{N^2 - p_1^2 - p_2^2}}. \tag{10.3}$$

On the other hand, $L = N\sqrt{1 + \dot{x}_1^2 + \dot{x}_2^2}$ and the Hamiltonian function is

$$H(x_\alpha, p_\alpha, x_3) = \sum_{\alpha=1}^{2} p_\alpha \dot{x}_\alpha - L = -\sqrt{N^2 - p_1^2 - p_2^2}. \quad (10.4)$$

Finally, in the present case, Hamilton's equations

$$\dot{x}_\alpha = \frac{\partial H}{\partial p_\alpha},$$

$$\dot{p}_\alpha = -\frac{\partial H}{\partial x_\alpha}$$

assume the following form:

$$\dot{x}_\alpha = \frac{p_\alpha}{\sqrt{N^2 - p_1^2 - p_2^2}}, \quad (10.5)$$

$$\dot{p}_\alpha = \frac{N}{\sqrt{N^2 - p_1^2 - p_2^2}} \frac{\partial N}{\partial x_\alpha}, \quad (10.6)$$

whereas the jump conditions (9.48) in the Hamiltonian formalism become

$$[[p_\alpha]] = \lambda n_\alpha. \quad (10.7)$$

Again we find that, provided the refractive index is constant, the rays are straight lines.

10.2 Hamilton's Principal Functions

This section is devoted to the **principal or characteristic functions** introduced by Hamilton to study an optical system. Knowledge of one of these functions allows us to perform a complete analysis of the aberrations in an optical system. Unfortunately, exact knowledge of them is not possible. However, this fact does not mean that these functions have no practical value. Indeed, it is possible to obtain approximate expressions for them. In particular, Gaussian optics is obtained from their first approximation. Moreover, their second approximation leads to third-order aberration theory, their third-order approximation to fifth-order aberration theory, and so on. However, the calculations needed to attain more accurate approximations of the Hamiltonian principal functions become so complex that only the first two approximations are actually useful.

10.2. Hamilton's Principal Functions

There are also other advantages of using the principal functions:

- We noted in Chapter 1 that Maupertuis' principle, which describes mechanical systems, is analogous to Fermat's principle in optics. Consequently, Jacobi proposed that the principal functions could also be applied in mechanics. This approach led to the famous Hamilton–Jacobi theory of canonical transformations, from which very important results were derived.

- Using the principal functions and Noether's theorem, it is easy to elucidate the implications of symmetries in an optical system.

We recall that the **object space** is a term used to denote the region (containing the object) that the optical system forms an image of. The light originates from this region, and by convention it is assumed to be situated on the left-hand side of the optical system. Similarly, the **image space** is the region containing the image formed by the optical system. It may coincide with the object space, for example in the reflecting systems we analyzed in previous chapters.

Practically speaking, it may be convenient to refer the object and image spaces using either the same coordinate system $Ox_1x_2x_3$ or with two different frames, $Ox_1x_2x_3$ and $Ox'_1x'_2x'_3$.

In order to introduce the first principal function, we suppose that:

For any pair of points **x** *and* **x**' *where the former occurs in the object space and the latter in the image space, there is one ray γ that has these points as end points.*

In other words, we are assuming that there are no pairs of stigmatic points in the above regions. This hypothesis implies that the optical path length $OPL(\gamma_0)$ along a ray depends only on the end points selected, since the ray γ_0 is uniquely determined by these points.

The **point principal function** is defined as follows:

$$V(x_i, x'_j) = OPL(\gamma_0), \tag{10.8}$$

where (x_i) and (x'_j) denote the coordinates of **x** and **x**', respectively, and *the optical path length is evaluated along the unique ray γ_0 that connects them.*

To illustrate the importance of the point principal function V, we need to evaluate its differential with respect to the points (x_i) and (x'_j). To this end, it is sufficient to take into account the one-parameter family Δ of curves with variable end points (see (9.11) and (9.12)), and to calculate the differential of the function $I(\epsilon)$ (see (9.10)) at $\epsilon = 0$; in other words, we need to use (9.31) *for a family of rays*. From (9.31), and taking (9.34) into

account, we have:
$$dV = \frac{\partial V}{\partial x_i}dx_i + \frac{\partial V}{\partial x'_i}dx'_i = -\frac{\partial L}{\partial \dot{x}_i}dx_i + \frac{\partial L}{\partial \dot{x}'_i}dx'_i. \quad (10.9)$$

In turn, owing to the arbitrary nature of the variations dx_i and dx'_i, this relation implies the equations
$$\frac{\partial V}{\partial x_i} = -\frac{\partial L}{\partial \dot{x}_i}, \quad \frac{\partial V}{\partial x'_i} = \frac{\partial L}{\partial \dot{x}'_i}, \quad (10.10)$$

which, due to (9.27) and (9.36), assume the form
$$p_i = -\frac{\partial V}{\partial x_i} = p_i(x_i, x'_i), \quad (10.11)$$
$$p'_i = \frac{\partial V}{\partial x'_i} = p'_i(x_i, x'_i), \quad (10.12)$$

where the usual notation
$$p_i = Nt_i$$
has been used. The system corresponding to (10.11) and (10.12) applies to any pair of points where one occurs in the object space and the other in the image space; the direction of the unique ray that connects them is $\mathbf{p}$ or $\mathbf{p}'$, respectively. We remark that (10.11) and (10.12) are not independent, since the following identities hold:
$$\mathbf{p} \cdot \mathbf{p} = N^2, \quad \mathbf{p}' \cdot \mathbf{p}' = N'^2, \quad (10.13)$$

which can also be written as follows due to (10.11) and (10.12):
$$\sum_{i=1}^{3}\left(\frac{\partial V}{\partial x_i}\right)^2 = N, \quad \sum_{i=1}^{3}\left(\frac{\partial V}{\partial x'_i}\right)^2 = N'. \quad (10.14)$$

We conclude that V is a solution to the eikonal equation in both the object and image spaces.

Suppose that the first two equations of (10.11) and the first two equations of (10.12),
$$p_\alpha = -\frac{\partial V}{\partial x_\alpha} = p_\alpha(x_\beta, x'_\beta, x_3, x'_3), \quad (10.15)$$
$$p'_\alpha = -\frac{\partial V}{\partial x'_\alpha} = p'_\alpha(x_\beta, x'_\beta, x_3, x'_3), \quad \alpha = 1,2, \quad (10.16)$$

form a system of four independent equations; in other words, that the following condition is satisfied:
$$\det\left(\frac{\partial^2 V}{\partial x_\alpha \partial x'_\beta}\right) \neq 0. \quad (10.17)$$

10.2. Hamilton's Principal Functions

In this hypothesis, the variables x'_β can be obtained from system (10.15). Upon substituting these variables into (10.16), the system corresponding to (10.15) and (10.16) assumes the form

$$x'_\alpha = x'_\alpha(x_\beta, p_\beta, x_3, x'_3), \tag{10.18}$$
$$p'_\alpha = p'_\alpha(x_\beta, p_\beta, x_3, x'_3). \tag{10.19}$$

Now let $x_3 = \bar{x}_3$ and $x'_3 = \bar{x}'_3$ be two planes in the object space and in the image space, respectively, which are assumed to have no pairs of stigmatic points. These planes are called the **anterior base plane** and the **posterior base plane**, respectively.

We can say that:

Equations (10.18) and (10.18) define a correspondence such that any particular point (x_α) in the anterior base plane $x_3 = \bar{x}_3$ and any particular ray direction p_α at this point is associated with only one point (x'_α) in the posterior base plane $x'_3 = \bar{x}'_3$ and only one ray direction at (x'_α).

We note that V cannot be used when the planes x_3 and x'_3 contain stigmatic points. If that happens, we can use other principal functions.

The **angular function** is defined as follows:

$$T = V + \mathbf{p} \cdot \mathbf{x} - \mathbf{p}' \cdot \mathbf{x}'. \tag{10.20}$$

If the media in the object and image spaces are homogeneous, a simple meaning can be attributed to the angular function T. Let r and r' be straight lines that come from O and O', respectively, and that are also perpendicular to the rays γ and γ' that emerge from A and A', respectively. Q and Q' are the intersection points of r and r' with γ and γ', respectively. Noting the meaning of V, we recognize that T coincides with the optical path length along the ray $QAA'Q'$ (see Figure 10.1)

From (10.9) and (10.20), we obtain

$$dT = \mathbf{x} \cdot d\mathbf{p} - \mathbf{x}' \cdot d\mathbf{p}', \tag{10.21}$$

where the variations $d\mathbf{p}$ and $d\mathbf{p}'$ are not arbitrary since the equations of (10.13) imply the conditions

$$\mathbf{p} \cdot d\mathbf{p} = 0, \quad \mathbf{p}' \cdot d\mathbf{p}' = 0. \tag{10.22}$$

Consequently,

$$dp_3 = -\frac{1}{p_3} \sum_{\alpha=1}^{2} p_\alpha dp_\alpha, \quad dp'_3 = -\frac{1}{p'_3} \sum_{\alpha=1}^{2} p'_\alpha dp'_\alpha. \tag{10.23}$$

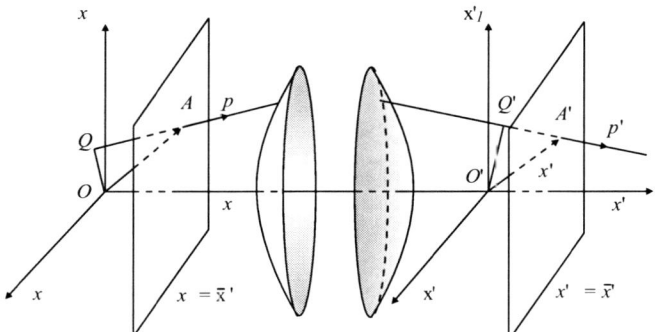

Fig. 10.1 Angular function

Substituting these expressions into (10.21), the following relation is obtained:

$$dT = \sum_{\alpha=1}^{2}\left[\left(x_\alpha - \frac{p_\alpha}{p_3}x_3\right)dp_\alpha - \left(x'_\alpha - \frac{p'_\alpha}{p'_3}x'_3\right)dp'_\alpha\right], \quad (10.24)$$

where the variations dp_α and dp'_α are now arbitrary. We then obtain

$$x_\alpha = \frac{p_\alpha}{p_3}x_3 + \frac{\partial T}{\partial p_\alpha}, \quad (10.25)$$

$$x'_\alpha = \frac{p'_\alpha}{p'_3}x'_3 - \frac{\partial T}{\partial p'_\alpha}. \quad (10.26)$$

Thus, if the angular function $T(p_\alpha, p'_\alpha)$ is known and x_3, x'_3 are fixed, (10.25) and (10.26) apply to any pair of ray directions **p**, **p**′ at the points **x** and **x**′ at which these rays intersect the anterior and posterior base planes.

Similarly, the **mixed functions** can be defined in the following way:

$$W_1 = V + \mathbf{p}\cdot\mathbf{x}, \quad W_2 = V - \mathbf{p}'\cdot\mathbf{x}', \quad (10.27)$$

where W_1 and W_2 denote the optical paths QA' and AQ', respectively.

From (10.20) and the above relations, we obtain

$$dW_1 = -\mathbf{x}\cdot d\mathbf{p} + \mathbf{p}'\cdot d\mathbf{x}', \quad (10.28)$$

$$dW_2 = -d\mathbf{x}\cdot\mathbf{p} - d\mathbf{p}'\cdot\mathbf{x}'. \quad (10.29)$$

The variations $d\mathbf{p}$ and $d\mathbf{p}'$ must satisfy the constraints of (10.23). Taking this into account and fixing the posterior base plane, we derive the following relations from (10.28) and (10.29):

$$x_\alpha = \frac{p_\alpha}{p_3} x_3 - \frac{\partial W_1}{\partial p_\alpha}, \qquad (10.30)$$

$$p'_\alpha = \frac{\partial W_1}{\partial x'_\alpha}. \qquad (10.31)$$

$$x'_\alpha = \frac{p'_\alpha}{p'_3} x'_3 + \frac{\partial W_2}{\partial p'_\alpha}, \qquad (10.32)$$

$$p_\alpha = -\frac{\partial W_2}{\partial x_\alpha}. \qquad (10.33)$$

The same approach as that applied to the system of (10.25) and (10.26) can be repeated for the system (10.30) and (10.31) and for the system (10.32) and (10.33). Finally, note that anomalous situations also exist for the functions T, W_1, and W_2. In fact, if **p** and **p'** are fixed and there are infinite pairs of points **x** and **x'** that satisfy (10.25) and (10.26) (telescopic systems), then the function T cannot be used. Similarly, if the system of (10.30) and (10.31) associates infinite pairs **p**, **x'** (focal point in the object plane) with a fixed pair **x**, **p'**, the function W_1 cannot be used. A similar situation holds for W_2.

10.3 Symmetries and Characteristic Functions

From now on, we will consider an optical system $\mathbb{S}$ with an axis of symmetry a, which is called the **optical axis** of the system. In this section, the restrictions on the characteristic functions derived from this axial symmetry are analyzed. Moreover, a further symmetry is considered for such a system; a symmetry with respect to a plane orthogonal to a (reversible system).

Let **Q** be any orthogonal matrix describing a rotation around a and $[\mathbf{Q}]_a$ be the set of these matrices. The axial symmetry mentioned above is then equivalent to the condition that any function $F(\mathbf{u}_1, \ldots, \mathbf{u}_n)$ of the vector variables $\mathbf{u}_1, \ldots, \mathbf{u}_n$ verifies the condition

$$F(\mathbf{u}_1, \ldots, \mathbf{u}_n) = F(\mathbf{Q}\mathbf{u}_1, \ldots, \mathbf{Q}\mathbf{u}_n), \quad \forall \mathbf{Q} \in [\mathbf{Q}]_s. \qquad (10.34)$$

If $\mathbf{u}_i^\perp$ is the component of the vector $\mathbf{u}_i$ that is orthogonal to a, and $\mathbf{u}_i^\parallel$ denotes the component that is parallel to s, it can be algebraically proven that (10.34) is satisfied if and only if

$$F(\mathbf{u}_1,\ldots,\mathbf{u}_n) = F(|\mathbf{u}_i^\perp|, \mathbf{u}_i^\perp \cdot \mathbf{u}_j^\perp, \mathbf{u}_i^\parallel). \tag{10.35}$$

For a system with axial symmetry, axis a is a ray for the optical system $\mathbb{S}$. In what follows, the axis Ox_3 of the reference frame $Ox_1x_2x_3$ in the object space will be chosen to coincide with a. In the image space, the same reference frame will be adopted or, equivalently, another frame Ox_1', x_2', x_3' where the axis Ox_3' coincides with Ox_3 and the other two axes are parallel to the corresponding axes of $Ox_1x_2x_3$ will be employed (see Figure 10.1). This choice of reference axes implies that the refractive index $N(\mathbf{x})$ must verify the condition

$$N(\mathbf{x}) = N(u, x_3), \quad u = x_1^2 + x_2^2. \tag{10.36}$$

Moreover, if $V(\mathbf{x}, \mathbf{x}')$ is the principal point function of an optical system $\mathbb{S}$ that exhibits axial symmetry around a, in the aforementioned coordinate system we have

$$V(x_1, x_2, x_3, x_1', x_2', x_3') = V(u, u', x_3, x_3'), \tag{10.37}$$

where

$$u = x_1^2 + x_2^2, \quad u' = {x_1'}^2 + {x_2'}^2, \quad w = x_1 x_1' - x_2 x_2'. \tag{10.38}$$

A similar approach can be applied to the other principal functions. For instance, for the angular function, we get

$$T(\mathbf{p}, \mathbf{p}') = T(\xi, \xi', \eta), \tag{10.39}$$

where

$$\xi = p_1^2 + p_2^2, \quad \xi' = {p_1'}^2 + {p_2'}^2, \quad \eta = p_1 p_1' + p_2 p_2'. \tag{10.40}$$

Note that p_3 and p_3' do not explicitly appear on the right-hand side of (10.39), since $p_3^2 = N^2 - (p_1^2 + p_2^2) = N^2 - \xi^2$, and a similar relation holds for p_3'.

An axially symmetric optical system $\mathbb{S}$ is said to be **reversible** if there is a plane of symmetry π that is orthogonal to a (see Figure 10.2).

Let γ be a ray that can be defined by the pair $\mathbf{p} = (p_1, p_2, p_3)$ and $\mathbf{p}' = (p_1', p_2', p_3')$. The mirror-like reflection of this ray with respect to π is another ray γ^* defined by $\mathbf{p}^* = (p_1, p_2, -p_3)$ in the image space (now interpreted as the object space), and by $\mathbf{p}^{*\prime} = (p_1^{*\prime}, p_2^{*\prime}, -p_3^{*\prime})$ in the object space (now regarded as the image space). Due to the reversibility of rays, γ can be traversed in the opposite direction, so the pair $\overline{\mathbf{p}} = -\mathbf{p}^{*\prime} = (-p_1', -p_2', -p_3')$

10.4. Lagrange's Optical Invariant for Axially Symmetric Systems   199

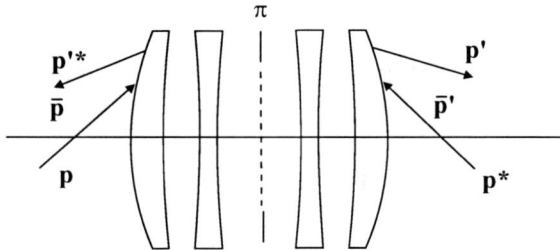

Fig. 10.2 Reversible optical system

again characterizes a ray $\overline{\mathbf{p}}' = -\mathbf{p}^* = (-p_1, -p_2, -p_3)$. By applying the reversibility condition to (10.39), we get

$$T(\xi, \xi', \eta) = T(\xi', \xi, \eta). \tag{10.41}$$

In other words, $T(\xi, \xi', \eta)$ is symmetric with respect to an exchange of variables (ξ and ξ').

10.4 Lagrange's Optical Invariant for Axially Symmetric Systems

Let $\mathbb{S}$ be an axially symmetric system with an optical axis a. Moreover,

$$\mathbf{x}^* = \phi(\epsilon, \mathbf{x}), \tag{10.42}$$

denotes the one-parameter group of rotations of $\mathbb{S}$ around a. Here, the parameter is the rotation angle ϵ. If the axes of the frames $Ox_1x_2x_3$ and $O'x'_1x'_2x'_3$ are chosen as in the previous section (i.e., with Ox_3 and $O'x'_3$ coinciding with a and the planes Ox_1x_2 and $O'x'_1x'_2$ orthogonal to a), the transformations (10.42) rotate the components of $\mathbf{x}$ and $\mathbf{x}'$ orthogonal to a, and do not modify the components parallel to a.

Due to the axial symmetry, the principal point function $V(\mathbf{x}, \mathbf{x}')$ must verify the condition

$$V(\mathbf{x}^\perp, \mathbf{x}'^\perp, \overline{x}_3, \overline{x}'_3) = V(\phi(\epsilon, \mathbf{x}^\perp), \phi(\epsilon, \mathbf{x}'^\perp), \overline{x}_3, \overline{x}'_3) \tag{10.43}$$

for any angle ϵ. Explicitly, the group of rotations (10.42) around a are expressed by the following relations:

$$x_1^* = x_1 \cos \epsilon + x_2 \sin \epsilon, \tag{10.44}$$
$$x_2^* = -x_1 \sin \epsilon + x_2 \cos \epsilon, \tag{10.45}$$

so that
$$\left(\frac{dx_1^*}{d\epsilon}\right)_{\epsilon=0} = x_2^*, \quad \left(\frac{dx_2^*}{d\epsilon}\right)_{\epsilon=0} = -x_1^*. \tag{10.46}$$

Differentiating (10.43) with respect to ϵ and evaluating the result at $\epsilon = 0$ yields the following relation:
$$\frac{\partial V}{\partial x_1}x_2 - \frac{\partial V}{\partial x_2}x_1 + \frac{\partial V}{\partial x_1'}x_2' - \frac{\partial V}{\partial x_2'}x_1' = 0. \tag{10.47}$$

Recalling (10.11) and (10.12), the above equation becomes
$$I \equiv p_2 x_1 - p_1 x_2 = p_2' x_1' - p_1' x_2'. \tag{10.48}$$

The quantity I is termed **Lagrange's optical invariant** since its value is constant upon traversing the whole optical system (see also Section 9.3 in order to account for discontinuities in the refractive index).

It is evident that this result is independent of the principal function used. For example, if the angular function T is used instead of (10.43), (10.44), (10.45), (10.46), and (10.48), we get the following equations:
$$T(\mathbf{p}^\perp, \mathbf{p}'^\perp) = T(\phi(\epsilon, \mathbf{p}^\perp), \phi(\epsilon, \mathbf{p}'^\perp)), \tag{10.49}$$

$$p_1^* = p_1 \cos\epsilon + p_2 \sin\epsilon, \tag{10.50}$$

$$p_2^* = -p_1 \sin\epsilon + p_2 \cos\epsilon, \tag{10.51}$$

$$\left(\frac{dp_1^*}{d\epsilon}\right)_{\epsilon=0} = p_2^*, \quad \left(\frac{dp_2^*}{d\epsilon}\right)_{\epsilon=0} = -p_1^*. \tag{10.52}$$

$$\frac{\partial T}{\partial p_1}p_2 - \frac{\partial T}{\partial p_2}p_1 + \frac{\partial T}{\partial p_1'}p_2' - \frac{\partial T}{\partial p_2'}p_1' = 0. \tag{10.53}$$

Then, because of (10.25) and (10.25), we again derive (10.48).

Chapter 11

Monochromatic Third-Order Aberrations

11.1 Introduction to Third-Order Aberrations

In Chapter 2 we presented the paraxial or Gaussian theory, which describes the propagation of light across an optical system $\mathbb{S}$, but only for rays that are close to the optical axis a and that subtend small angles with it. We showed that, under these conditions, the correspondence between an *assigned pair of object and image planes* π and π' is given by the linear relations

$$x' = Mx, \qquad (11.1)$$
$$y' = My, \qquad (11.2)$$

where M is the magnification of $\mathbb{S}$.

The aim of the optical design is usually to extend the paraxial correspondence described by (11.1) and (11.2) to include points as far from the optical axis a as possible, as well as rays that subtend greater angles with a than those considered in the Gaussian approximation.

We recall that a system $\mathbb{S}$ is said to exhibit **aberrations** if the effective correspondence

$$x' = x'(x, y, p, q), \qquad (11.3)$$
$$y' = y'(x, y, p, q), \qquad (11.4)$$

between the object and image planes π and π' differs from that given by (11.1) and (11.2). Therefore, the **aberration vector** ϵ is defined as the vector whose components correspond to the differences (see Section 1.4)

$$\epsilon_x = x' - Mx, \qquad (11.5)$$
$$\epsilon_y = y' - My. \qquad (11.6)$$

The aim of aberration theory is to derive the analytic expressions of (11.5) and (11.6) in terms of the coordinates of the object, the ray, and the characteristics of the optical system $\mathbb{S}$. These expressions are then used to both evaluate the aberrations of a given optical system $\mathbb{S}$ and to design $\mathbb{S}$ in such a way that the components of the aberration vector ϵ have pre-fixed values upon varying both the coordinates of the object point and the ray over a given range.

We have already noted that Taylor's expansion of the functions (11.5) and (11.6) with respect to the variables x, y, p, q only contains odd terms due to the symmetry of revolution of $\mathbb{S}$ about the optical axis a. The first-order terms define the Gaussian correspondence between π and π', the third-order terms express the **third-order aberrations**, the fifth-order terms describe the **fifth-order aberrations**, and so on. Correspondingly, the system $\mathbb{S}$ is said to be free from third-order aberrations if the third-order terms are absent; corrected up to the fifth order if the fifth-order aberrations vanish; etc.

In this chapter, formulae expressing the third-order aberrations of an optical system for an assigned pair of object and image planes are derived explicitly using the Hamiltonian approach. The corresponding formulae for the fifth-order aberrations are not given here because their complexity makes them difficult to apply (see [5, 6]).

11.2 Third-Order Aberrations via the Angular Function

First, we note that if the refractive index N of the medium before the first surface of the optical system is equal to the refractive index N' after the last surface, then, from (2.78), (2.79), the Gaussian transformations can be written as follows:

$$x' = Mx,$$
$$y' = My,$$
$$p' = -\frac{N}{f}x + \frac{1}{M}p, \qquad (11.7)$$
$$q' = -\frac{N}{f}y + \frac{1}{M}q.$$

On the other hand, considering (10.39) and (10.40), the characteristic angular function $T(p, q, p', q')$ is (to within sixth-order terms)

$$T = T_0 + T_2 + T_4, \qquad (11.8)$$

11.2. Third-Order Aberrations via the Angular Function

where T_0 is a constant,

$$T_2 = A_{\xi}\xi + A_{\xi'}\xi' + A_{\eta}, \tag{11.9}$$

and

$$T_4 = A_{\xi\xi}\xi^2 + A_{\xi'\xi'}\xi'^2 + A_{\eta\eta}\eta^2 + A_{\xi\eta}\xi\eta + A_{\xi\xi'}\xi\xi' + A_{\xi'\eta}\xi'\eta. \tag{11.10}$$

Now we suppose that the anterior and posterior base planes coincide, respectively, with the object plane Oxy and the image plane $O'x'y'$. Consequently, in the formulae (10.25) and (10.26) we have $x_3 = x'_3 = 0$. Components (11.5) and (11.6) of the aberration vector can then be written in the following way:

$$\epsilon_x = x' - Mx = -\frac{\partial T}{\partial p'} - M\frac{\partial T}{\partial p}, \tag{11.11}$$

$$\epsilon_y = y' - My = -\frac{\partial T}{\partial q'} - M\frac{\partial T}{\partial q}. \tag{11.12}$$

On the other hand, in the Gaussian approximation we have $x' - Mx = 0$ and $y' - My = 0$. Consequently,

$$-\frac{\partial T_2}{\partial p'} - M\frac{\partial T_2}{\partial p} = 0, \tag{11.13}$$

$$-\frac{\partial T_2}{\partial q'} - M\frac{\partial T_2}{\partial q} = 0, \tag{11.14}$$

and relations (11.11) and (11.12) assume the following form:

$$\epsilon_x = -\frac{\partial T_4}{\partial p'} - M\frac{\partial T_4}{\partial p}, \tag{11.15}$$

$$\epsilon_y = -\frac{\partial T_4}{\partial q'} - M\frac{\partial T_4}{\partial q}. \tag{11.16}$$

We thus conclude that the third-order aberrations depend only on T_4.

If (11.10) is taken into account, these equations become

$$\epsilon_x = -\mathbb{A}p - \mathbb{B}p', \tag{11.17}$$

$$\epsilon_y = -\mathbb{A}q - \mathbb{B}q', \tag{11.18}$$

where we have introduced the following notation:

$$\mathbb{A} = 2M(2A_{\xi\xi}\xi + A_{\xi\xi'}\xi' + A_{\xi\eta}\eta)$$
$$+ (A_{\xi\eta}\xi + A_{\xi'\eta}\xi' + 2A_{\eta\eta}\eta), \tag{11.19}$$

$$\mathbb{B} = M(A_{\xi\eta}\xi + A_{\xi'\eta}\xi' + 2A_{\eta\eta}\eta)$$
$$+ 2(A_{\xi\xi'}\xi + 2A_{\xi'\xi'}\xi' + A_{\xi'\eta}\eta). \tag{11.20}$$

These relations do not appear to be useful. Our aim is to express the aberrations in terms of both the coordinates x, y of the object point and the directional numbers p, q of the emerging ray. In contrast, together with p, q, the directional numbers p', q' of the corresponding ray in the image space appear instead of x, y in (11.19) and (11.20). Fortunately, this difficulty is easily overcome. Since we are interested in relations that are exact to within third-order terms, it is sufficient to resort to the *linear relations* of (11.7), which express p' and q' in terms of x, y, p, q. A further simplification can also be introduced. Due to the symmetry of the optical system $\mathbb{S}$, it is possible to assume that $x = 0$ without any loss of generality. In other words, the following equations can be used:

$$p' = \frac{1}{M}p, \tag{11.21}$$

$$q' = -\frac{N}{f}y + \frac{1}{M}q. \tag{11.22}$$

In what follows, it will be more convenient to describe the ray emerging from the object point $(0, y)$ using the *paraxial* coordinates x_e, y_e of the point at which it crosses the plane of the entrance pupil. Let P_e, P_e', and P' be the points at which a ray γ from an object point P intersects the plane of the entrance pupil π_e, the plane of the exit pupil π_e', and the conjugate Gaussian object plane π', respectively. The parametric equations for the ray γ emerging from $(0, y)$ are

$$X = pt, \tag{11.23}$$
$$Y = y + qt, \tag{11.24}$$
$$Z = rt, \tag{11.25}$$

where t is the parameter and (X, Y, Z) is the moving point on γ. Since we are interested in relations that are linear with respect to the variables p, q, r, y, we have

$$Z \simeq Nt, \tag{11.26}$$

so that the value $\bar{t}$ corresponding to the point P_e is given by

$$\bar{t} = \frac{z_e}{N}, \tag{11.27}$$

where z_e denotes the abscissa of π_e with respect to the object plane. Substituting this value of t into the other two equations (11.24) and (11.25) leads to

$$p = \frac{N}{z_e}x_e, \tag{11.28}$$

$$q = \frac{N}{z_e}(y_e - y). \tag{11.29}$$

11.2. Third-Order Aberrations via the Angular Function

After using (11.28) and (11.29) in (11.17) and (11.18), and tedious algebraic manipulation, we derive the following expressions for the components of the aberration vector:

$$\epsilon_x = \mathbb{D}_1(x_e^2 + y_e^2)x_e + 2\mathbb{D}_2 x_e y_e y + \mathbb{D}_3 x_e y^2, \tag{11.30}$$

$$\epsilon_y = \mathbb{D}_1(x_e^2 + y_e^2)y_e + \mathbb{D}_2 x_e^2 y + 3\mathbb{D}_2 y_e^2 y \\ + \mathbb{D}_4 y_e y^2 + \mathbb{D}_5 y^3, \tag{11.31}$$

where the following notation has been introduced:

$$\mathbb{D}_1 = -\frac{4N^3}{z_e^3}A_1,$$

$$\mathbb{D}_2 = \frac{N^3}{z_e^3}\left(4A_1 + \frac{z_e}{f}A_2\right),$$

$$\mathbb{D}_3 = -\frac{2N^3}{z_e^3}\left(2A_1 + \frac{z_e}{f}A_2 + \frac{z_e^2}{f^2}A_3\right),$$

$$\mathbb{D}_4 = -\frac{2N^3}{z_e^3}\left(6A_1 + 3\frac{z_e}{f}A_2 + \frac{z_e^2}{f^2}A_4\right),$$

$$\mathbb{D}_5 = \frac{N^3}{z_e^3}\left(4A_1 + 3\frac{z_e}{f}A_2 + 2\frac{z_e^2}{f^2}A_4 + \frac{z_e^3}{f^3}A_5\right),$$

$$A_1 = MA_{\xi\xi} + \frac{A_{\xi\xi'}}{M} + A_{\xi\eta} + \frac{A_{\xi'\xi'}}{M^3} + \frac{A_{\xi'\eta}}{M^2} + \frac{A_{\eta\eta}}{M},$$

$$A_2 = 2A_{\xi\xi'} + MA_{\xi\eta} + 4\frac{A_{\xi'\xi'}}{M^2} + \frac{3A_{\xi'\eta}}{M} + 2A_{\eta\eta},$$

$$A_3 = MA_{\xi\xi'} + 2\frac{A_{\xi'\xi'}}{M} + A_{\xi'\eta},$$

$$A_4 = MA_{\xi\xi'} + 6\frac{A_{\xi'\xi'}}{M} + 3A_{\xi'\eta} + MA_{\eta\eta},$$

$$A_5 = 4A_{\xi'\xi'} + MA_{\xi'\eta}.$$

In Chapter 3, we discussed the meaning of the terms that appear in the above equations.

We have now derived many important properties of the third-order aberrations using elementary tools. However, there are still two important tasks to perform:

- Determine, for a *single* optical surface, the fourth-order expansion of the angular function in terms of its optical characteristics

- Evaluate the *total* angular function in order to derive the coefficients $A_1, \ldots, A_5$ that appear in the above formulae.

We tackle these tasks in the next few sections.

11.3 Reduced Coordinates

In mechanics, it is very common to find a problem that cannot be solved in its original formulation. However, selecting suitable Lagrangian or Hamiltonian coordinates can result in a reformulated equation of motion that may allow us to find the solution or, at the very least, more information about it. That also happens when studying optical systems. In this section, we introduce the **reduced coordinates** of Seidel, since they allow us to formulate the aberration problem in a simpler form.

Let $\mathbb{S}$ be an axially symmetric optical system, a be its axis of symmetry, and π and π' be a pair of Gaussian object and image planes for $\mathbb{S}$. We introduce two reference lengths l and l' in π and π', respectively, which verify the condition

$$\frac{l'}{l} = M, \tag{11.32}$$

where M is the lateral Gaussian magnification related to π and π'.

In these planes, we will use the following nondimensional coordinates:

$$X = C\frac{x}{l}, \qquad Y = C\frac{y}{l}, \tag{11.33}$$

$$X' = C\frac{x'}{l'}, \qquad Y' = C\frac{y'}{l'}, \tag{11.34}$$

where C is a constant that will be fixed later.

Similarly, we introduce another two reference lengths l_e and l'_e in the conjugate planes of the entrance and exit pupils such that

$$\frac{l'_e}{l_e} = M_e, \tag{11.35}$$

where M_e is the magnification related to these two planes.

Correspondingly, we use the following nondimensional coordinates for these planes:

$$X_e = \frac{x_e}{l_e}, \qquad Y_e = \frac{y_e}{l_e}, \tag{11.36}$$

$$X'_e = \frac{x'_e}{l'_e}, \qquad Y'_e = \frac{y'_e}{l'_e}. \tag{11.37}$$

If z_e and z'_e denote, respectively, the distances of the entrance and exit pupils from π and π' in the Gaussian approximation (see (11.28) and (11.29)), we find that

$$\frac{x_e - x}{z_e} = \frac{p}{N}, \qquad \frac{y_e - y}{z_e} = \frac{q}{N} \tag{11.38}$$

$$\frac{x'_e - x'}{z'_e} = \frac{p'}{N'}, \qquad \frac{y'_e - y'}{z'_e} = \frac{q'}{N'}, \tag{11.39}$$

11.4. Schwarzschild's Eikonal

and the relations (11.36) and (11.37) can also be written as

$$X_e = \frac{x}{l_e} + \frac{z_e p}{l_e N}, \qquad Y_e = \frac{y}{l_e} + \frac{z_e q}{l_e N}, \tag{11.40}$$

$$X'_e = \frac{x'}{l'_e} + \frac{z'_e p'}{l'_e N'}, \qquad Y'_e = \frac{y'}{l'_e} + \frac{z'_e q'}{l'_e N'}. \tag{11.41}$$

It is important to note that in the Gaussian approximation we have

$$X' = X, \qquad Y' = Y, \tag{11.42}$$
$$X'_e = X_e, \qquad Y'_e = Y_e. \tag{11.43}$$

Consequently, the differences $X' - X$, $Y' - Y$, $X'_e - X_e$, and $Y'_e - Y_e$ express the aberrations in the optical system $\mathbb{S}$.

In what follows, we choose the constant C in the following way in order to simplify the calculations:

$$C = \frac{N l l_e}{z_e} = \frac{N' l' l'_e}{z'_e}. \tag{11.44}$$

This choice is available to us because the ratios that appear in (11.44) coincide with Herschel's invariant (2.48).

The quantities defined by (11.33), (11.34), (11.40), and (11.41) are called the **Seidel variables**. We conclude by writing the inverse relations of (11.33) and (11.34) and of (11.40) and (11.41) when (11.44) is chosen:

$$x = \frac{z_e}{N l_e} X, \qquad y = \frac{z_e}{N l_e} Y, \tag{11.45}$$

$$x' = \frac{z'_e}{N' l'_e} X', \qquad y' = \frac{z'_e}{N' l'_e} Y', \tag{11.46}$$

$$p = \frac{N l_e}{z_e} X_e - \frac{1}{l_e} X, \qquad q = \frac{N l_e}{z_e} Y_e - \frac{1}{l_e} Y, \tag{11.47}$$

$$p' = \frac{N' l'_e}{z'_e} X'_e - \frac{1}{l'_e} X', \qquad q' = \frac{N' l'_e}{z'_e} Y'_e - \frac{1}{l'_e} Y'. \tag{11.48}$$

11.4 Schwarzschild's Eikonal

We start by noting that relation (10.24), in the coordinates adopted here, is written as

$$dT = x\,dp + y\,dq - x'\,dp' - y'\,dq'. \tag{11.49}$$

Introducing the Seidel variables (11.45)–(11.48), we obtain

$$dT = x\left(\frac{N l_e}{z_e} dX_e - \frac{1}{l_e} dX\right) + y\left(\frac{N l_e}{z_e} dY_e - \frac{1}{l_e} dY\right)$$

208   Chapter 11. Monochromatic Third-Order Aberrations

$$-x'\left(\frac{N'l'_e}{z'_e}dX'_\epsilon - \frac{1}{l'_e}dX'\right) - y'\left(\frac{N'l'_e}{z'_e}dY'_\epsilon - \frac{1}{l'_e}dY'\right),$$

in other words

$$dT = -\frac{z_e}{Nl_e^2}(XdX + YdY) + \frac{z'_e}{N'l_e'^2}(X'dX' + Y'dY')$$
$$+ XdX_e + YdY_e - X'dX'_e - Y'dY'_e.$$

It is easy to verify that the above expression is equivalent to the following:

$$dT = -\frac{z_e}{Nl_e^2}(XdX + YdY) + \frac{z'_e}{N'l_e'^2}(X'dX' + Y'dY')$$
$$- (X'_e - X_e)dX - (Y'_e - Y_e)dY - X(dX'_e - dX_e) - Y(dY'_e - dY_e)$$
$$+ (X'_e - X_e)dX + (Y'_e - Y_e)dY - (X' - X)dX'_e - (Y' - Y)dY'_e.$$

By introducing the function

$$\psi = \frac{z_e}{2Nl_e^2}(X^2 + Y^2) - \frac{z'_e}{2N'l_e'^2}(X'^2 + Y'^2)$$
$$+ (X'_e - X_e)X + (Y'_e - Y_e)Y, \qquad (11.50)$$

the above relation becomes

$$d(T + \psi) = (X'_e - X_e)dX + (Y'_e - Y_e)dY - (X' - X)dX'_e - (Y' - Y)dY'_e. \qquad (11.51)$$

In other words, $d(T + \psi)$ is the differential of the function

$$S = T + \psi, \qquad (11.52)$$

which is called the **Schwarzschild eikonal**.

In view of (11.51)–(11.52), it can be said that Schwarzschild's eikonal, when regarded as a function of Seidel's variables X, Y, X'_e, Y'_e, verifies the following important relations:

$$X'_e - X_e = \frac{\partial S}{\partial X}, \qquad Y'_e - Y_e = \frac{\partial S}{\partial Y}, \qquad (11.53)$$

$$X' - X = -\frac{\partial S}{\partial X'_e}, \qquad Y' - Y = -\frac{\partial S}{\partial Y'_e}. \qquad (11.54)$$

Remark We have already been noted that the following applies in the Gaussian approximation:

$$X' = X, \quad Y' = Y, \quad X'_e = X_e, \quad Y'_e = Y_e. \qquad (11.55)$$

11.4. Schwarzschild's Eikonal

These relations as well as (11.53) and (11.54) imply that Taylor's power expansion of S *does not contain second-order terms*; in other words, it takes the form

$$S = S^{(0)} + S^{(4)} + S^{(6)} + \cdots, \qquad (11.56)$$

where $S^{(2k)}$ is a homogeneous polynomial of degree $2k$ in Seidel's variables.

Remark From (11.53), we obtain

$$X' = X + O(3), \quad Y' = Y + O(3), \qquad (11.57)$$
$$X'_e = X_e + O(3), \quad Y'_e = Y_e + O(3). \qquad (11.58)$$

Moreover, due to the axial symmetry of the optical system $\mathbb{S}$, the Schwarzschild function S depends on the variables X, Y, X'_e, Y'_e through the following three combinations:

$$\mathcal{R}^2 = X^2 + Y^2, \quad \mathcal{R}'^2_e = X'^2_e + Y'^2_e, \quad k = XX'_e + YY'_e. \qquad (11.59)$$

In particular, the fourth-order term in (11.56) is

$$S^{(4)} = -\frac{1}{4}A\mathcal{R}^4 - \frac{1}{4}B\mathcal{R}'^4_e - Ck^4 - \frac{1}{2}D\mathcal{R}^2\mathcal{R}'^2_e + E\mathcal{R}^2k^2 + F\mathcal{R}'^2_e k^2. \qquad (11.60)$$

In order to evaluate the third-order aberrations from $S^{(4)}$, we start by noting that (11.33), (11.34), and (11.44) provide us with the relations

$$X' - X = \frac{N'l'_e}{z'_e}(x' - Mx), \quad Y' - Y = \frac{N'l'_e}{z'_e}(y' - My), \qquad (11.61)$$

which, considering (11.5) and (11.6), can also be written as follows:

$$X' - X = \frac{N'l'_e}{z'_e}\epsilon_x \equiv \alpha\epsilon_x, \quad Y' - Y = \frac{N'l'_e}{z'_e}\epsilon_y \equiv \alpha\epsilon_y. \qquad (11.62)$$

Finally, (11.53) and (11.54) lead us to the following relations:

$$\alpha\epsilon_x = (2Ck^2 - E\mathcal{R}^2 - F\mathcal{R}'^2_e)X + (B\mathcal{R}'^2_e + D\mathcal{R}^2 - 2Fk^2)X'_e,$$
$$\alpha\epsilon_y = (2Ck^2 - E\mathcal{R}^2 - F\mathcal{R}'^2_e)Y + (B\mathcal{R}'^2_e + D\mathcal{R}^2 - 2Fk^2)Y'_e.$$

If we assume $X = 0$ due to the axial symmetry of the optical system $\mathbb{S}$, and introduce the polar coordinates

$$X'_e = \mathcal{R}'_e \cos\varphi, \quad Y'_e = \mathcal{R}'_e \sin\varphi,$$

in the exit pupil, the above relations become

$$\alpha\epsilon_x = B\mathcal{R}'^3_e \cos\varphi - FY\mathcal{R}'^2_e \sin 2\varphi + DY^2\mathcal{R}'_e \cos\varphi, \qquad (11.63)$$
$$\alpha\epsilon_y = B\mathcal{R}'^3_e \sin\varphi - FY\mathcal{R}'^2_e(2 - \cos 2\varphi)$$
$$+ (2C + D)Y^2\mathcal{R}'_e \sin\varphi - EY^3. \qquad (11.64)$$

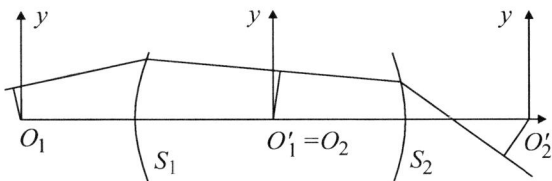

Fig. 11.1 Addition of angular functions in a compound system

Starting from these equations, which have the same form as (11.30) and (11.31), we can perform the same analysis of aberrations as that described in Section 3.6.

We conclude this section by noting that, since ψ is a second-order function of Seidel's variables and (11.52) holds, then

$$S^{(4)} = T^{(4)}, \tag{11.65}$$

where $T^{(4)}$ is the fourth-order term of Taylor's expansion of the angular function.

11.5 Addition Theorem for Third-Order Aberrations

In this section we demonstrate the great advantage offered by Seidel's variables when evaluating the third-order aberrations of a compound system.

Consider an axially symmetric optical system $\mathbb{S}$ consisting of two surfaces of revolution S_1 and S_2 (see Figure 11.1).

Let O_1 be the axial object point for the whole system, O_1' be the Gaussian image formed by S_1, $O_2 \equiv O_1'$ be the object of S_2, and O_2' be the image formed by S_2. If

$$T_1 = T_1(p_1, q_1, p_1', q_1'), \qquad T_2 = T_2(p_2, q_2, p_2', q_2'),$$

denote the angular functions related to S_1 and S_2 respectively, the angular function of the whole system

$$T = T_1 + T_2.$$

We need to eliminate the variables $p_1' = p_2$ and $q_1' = q_2$ from this expression. According to (11.50) and (11.52), Schwarzschild's eikonals of S_1 and

11.5. Addition Theorem for Third-Order Aberrations

S_2 are

$$S_1 = T_1 + \frac{z_{e1}}{2N_1 l_{e1}^2}(X_1^2 + Y_1^2) - \frac{z'_{e1}}{N'_1 l'^2_{e1}}(X_1'^2 + Y_1'^2)$$
$$+ (X'_{e1} - X_{e1})X_1 + (Y'_{e1} - Y_{e1})Y_1, \qquad (11.66)$$

$$S_2 = T_2 + \frac{z_{e2}}{2N_2 l_{e2}^2}(X_2^2 + Y_2^2) - \frac{z'_{e2}}{N'_2 l'^2_{e2}}(X_2'^2 + Y_2'^2)$$
$$+ (X'_{e2} - X_{e2})X_2 + (Y'_{e2} - Y_{e2})Y_2, \qquad (11.67)$$

whereas Schwarzschild's eikonal for the whole system is

$$S = T + \frac{z_{e1}}{2N_1 l_{e1}^2}(X_1^2 + Y_1^2) - \frac{z'_{e2}}{N'_2 l'^2_{e2}}(X_2'^2 + Y_2'^2)$$
$$+ (X'_{e2} - X_{e1})X_1 + (Y'_{e2} - Y_{e1})Y_1. \qquad (11.68)$$

Since $T = T_1 + T_2$, if we recall that $z'_{e1} = z_{e2}$ and $l'_{e1} = l_{e2}$, we easily obtain

$$S = S_1 + S_2 + (X_1 - X'_1)(X'_{e2} - X_{e2}) + (Y_1 - Y'_1)(Y'_{e2} - Y_{e2}). \qquad (11.69)$$

Using (11.53) and (11.54), we finally get

$$S = S_1 + S_2 + \frac{\partial S_1}{\partial X'_{e1}} \frac{\partial S_2}{\partial X_2} + \frac{\partial S_1}{\partial Y'_{e1}} \frac{\partial S_2}{\partial Y_2}, \qquad (11.70)$$

where, as usual, S_i is a function of X_i, Y_i and X'_{ei}, Y'_{ei}.

If the power expansions for S_1 and S_2 are introduced into (11.70), then, due to the absence of second-order terms in these expansions and taking into account (11.53) and (11.54), it follows that

$$S = S_1^{(0)} + S_2^{(0)} + S_1^{(4)} + S_2^{(4)} + \cdots. \qquad (11.71)$$

According to (11.60), $S_1^{(4)}$ and $S_2^{(4)}$ can be written as follows:

$$S_1 = -\frac{1}{4}A_1 \mathcal{R}_1^4 - \frac{1}{4}B_1 \mathcal{R}'^4_{e1} - C_1 k_1^4$$
$$- \frac{1}{2}D_1 \mathcal{R}_1^2 \mathcal{R}^2_{e1} + E_1 \mathcal{R}_1^2 k_1^2 + F_1 \mathcal{R}^2_{e1} k_1^2, \qquad (11.72)$$

$$S_2 = -\frac{1}{4}A_2 \mathcal{R}_2^4 - \frac{1}{4}B_2 \mathcal{R}'^4_{e2} - C_2 k_2^4$$
$$- \frac{1}{2}D_2 \mathcal{R}_2^2 \mathcal{R}^2_{e2} + E_2 \mathcal{R}_2^2 k_2^2 + F_2 \mathcal{R}^2_{e2} k_2^2, \qquad (11.73)$$

where

$$\mathcal{R}_1^2 = X_1^2 + Y_1^2, \quad \mathcal{R}'^2_{e1} = X'^2_{e1} + Y'^2_{e1}, \quad k_1 = X_1 X'_{e1} + Y_1 Y'_{e1},$$
$$\mathcal{R}_2^2 = X_2^2 + Y_2^2, \quad \mathcal{R}'^2_{e2} = X'^2_{e2} + Y'^2_{e2}, \quad k_2 = X_2 X'_{e2} + Y_2 Y'_{e2}.$$

212   Chapter 11. Monochromatic Third-Order Aberrations

Using (11.54), we can then easily derive

$$S = -\tfrac{1}{4}(A_1 + A_2)\mathcal{R}_1^4 - \tfrac{1}{4}(B_1 + B_2)\mathcal{R}_{e2}^{\prime 4} - (C_1 + C_2)k_{12}^4 \\ -\frac{1}{2}(D_1 + D_2)\mathcal{R}_1^2\mathcal{R}_{e2}^2 + (E_1 + E_2)\mathcal{R}_1^2 k_{12}^2 + (F_1 + F_2)\mathcal{R}_{e1}^2 k_{12}^2, \quad (11.74)$$

where

$$k_{12} = X_1 X_{e2}' + Y_1 Y_{e2}'.$$

A similar result holds for a centered system consisting of any number of surfaces. Thus, we have proved the following theorem.

Theorem 11.1
Each third-order aberration coefficient of a centered system is the sum of the corresponding coefficients of the single surfaces of the system.

11.6 Fourth-Order Expansion of the Angular Function

In this section, the angular characteristic function T of a single refractive or reflecting revolving surface S will be determined up to fourth-order terms in ξ, ξ', η.

Recall that the general expression (10.20) for the angular characteristic function (see also Figure 11.1) is

$$T = (p - p')x + (q - q')y + (r - r')z + r'a' - ra, \quad (11.75)$$

where $P = (x, y, z)$ is the point of intersection between S and the ray **p**. As a consequence, its coordinates must minimize the optical path length; in other words, the function

$$T(x, y, z) + \lambda F(x, y, z), \quad (11.76)$$

where λ is a Lagrangian multiplier and $F(x, y, z) = z - z(x, y) = 0$ is the equation of S. The conditions required for a minimum are

$$\frac{\partial T}{\partial x} + \lambda \frac{\partial F}{\partial x} = 0, \quad \frac{\partial T}{\partial y} + \lambda \frac{\partial F}{\partial y} = 0, \quad \frac{\partial T}{\partial z} + \lambda \frac{\partial F}{\partial z} = 0,$$

or, equivalently,

$$p - p' = \lambda \frac{\partial z}{\partial x}, \quad q - q' = \lambda \frac{\partial z}{\partial y}, \quad r - r' = -\lambda. \quad (11.77)$$

11.6. Fourth-Order Expansion of the Angular Function

Substituting the value of λ derived from the third of these equations into the others yields

$$\frac{p-p'}{r-r'} = -\frac{\partial z}{\partial x}, \quad \frac{q-q'}{r-r'} = -\frac{\partial z}{\partial y}. \tag{11.78}$$

We also recall (see (1.25)) that the equation of the surface S is (up to fourth-order terms):

$$z = \frac{(x^2+y^2)}{2R} + \frac{(1+K)}{8R^3}(x^2+y^2)^2. \tag{11.79}$$

Consequently, relations (11.78) become

$$\frac{p-p'}{r-r'} = -\frac{x}{R} + \Delta x, \quad \frac{q-q'}{r-r'} = -\frac{y}{R} + \Delta y, \tag{11.80}$$

where Δx and Δy are third-order quantities.

Moreover,

$$r = N - \frac{1}{2N}(p^2+q^2) + O(4), \quad r' = -\frac{1}{2N'}(p'^2+q'^2) + O(4), \tag{11.81}$$

and

$$\frac{1}{r'-r} = \frac{1}{N'-N}\left[1 - \frac{1}{2N(N'-N)}(p^2+q^2)\right.$$

$$\left. + \frac{1}{2N(N'-N)}(p'^2+q'^2)\right]. \tag{11.82}$$

Inserting (11.80) into (11.79) gives

$$z = \frac{R}{2(r-r')}[(p-p')^2 + (q-q')^2] - \frac{1}{r-r'}[(p-p')\Delta x + (q-q')\Delta y]$$

$$+ R\frac{1+K}{(r-r')^4}[(p-p')^2 + (q-q')^2]^2. \tag{11.83}$$

Taking into account this relation, (11.75) writes

$$T = -\frac{R}{r-r'}\left[(p-p')^2 + (q-q')^2\right.$$
$$+ (p-p')\Delta x + (q-q')\Delta y]$$
$$+ (r-r')\frac{R}{2(r-r')^2}\left[(p-p')^2 + (q-q')^2\right]$$
$$+ \frac{r-r'}{r-r'}(p-p')\Delta x + (q-q')\Delta y$$
$$+ R\frac{(r-r')(1+K)}{8(r-r')^4}\left[(p-p')^2 + (q-q')^2\right]^2$$

$$-\frac{R}{r-r'}\left[(p-p')^2+(q-q')^2\right]$$
$$+R\frac{((1+K))}{8(r-r')^3}\left[(p-p')^2+(q-q')^2\right]^2, \tag{11.84}$$

and the third-order terms Δx and Δy disappear.

The final expression of T is then

$$T = T_0 + T_2 + T_4, \tag{11.85}$$

where

$$T_0 = N'a' - Na, \tag{11.86}$$

$$T_2 = \frac{R}{2(N'-N)}\left[(p-p')^2+(q-q')^2\right]$$
$$+\frac{a}{2N}(p^2+q^2) - \frac{a'}{2N'}(p'^2+q'^2), \tag{11.87}$$

$$T_4 = -\frac{R}{4(N'-N)^2}\left[(p-p')^2+(q-q')^2\right]\left[\frac{p^2+q^2}{N^2}-\frac{p'^2+q'^2}{N'}\right]$$
$$-\frac{R(1+K)}{8(N'-N)^3}\left[(p-p')^2+(q-q')^2\right]^2$$
$$+\frac{N}{8N^3}(p^2+q^2)^2 - \frac{N'}{8N'^3}(p'^2+q'^2)^2. \tag{11.88}$$

Now the axial points $z = a$ and $z = a'$ are identified as the axial object point and its Gaussian image, respectively, and the following notation is adopted:

$$z = a, \quad z' = a', \quad w_e = a + z_e, \quad w'_e = a^{\cdot} + z'_e, \tag{11.89}$$

and

$$Q \equiv N\left(\frac{1}{R}-\frac{1}{z}\right) = N'\left(\frac{1}{R}-\frac{1}{z'}\right),$$
$$L \equiv N\left(\frac{1}{R}-\frac{1}{w_e}\right) = N'\left(\frac{1}{R}-\frac{1}{w'_e}\right), \tag{11.90}$$

where Q and L denote Abbe's invariants of the surface S for the object–image planes and entrance–exit pupils (see (2.7)).

Before expressing T_4 in terms of Seidel's variables, it is convenient to rewrite T_4 in (11.88) in a more suitable form. Introducing the notation of (11.89) into (11.88), we get

$$-\frac{R(1+K)}{8(N'-N)^3}\left[(p-p')^2+(q-q')^2\right]^2$$
$$= -\frac{R^2(1+K)}{8(N'-N)^4}\left[\frac{N'}{z'}-\frac{N}{z}\right]\left[(p-p')^2+(q-q')^2\right]^2,$$

and (11.88) assumes the following form:

$$T_4 = \frac{1}{8Nz}\left[\frac{NR}{(N'-N)^2}[(p-p')^2+(q-q')^2] - \frac{z}{N}(p^2+q^2)\right]$$
$$- \frac{1}{8N'z'}\left[\frac{N'R}{(N'-N)^2}[(p-p')^2+(q-q')^2] - \frac{z'}{N'}(p'^2+q'^2)\right]$$
$$- \frac{KR}{8(N'-N)^3}[(p-p')^2+(q-q')^2]. \qquad (11.91)$$

11.7 Aberrations of Axially Symmetric Systems

In order to find the third-order aberrations of an axially symmetric system, we must:

1. Express T_4 in terms of Seidel's variables
2. Apply the theorem for the addition of primary aberrations.

In relation to the first point, we start by noting that formulae (11.54) allow us to use the Gaussian relations (11.61) in (11.91). Moreover, since Seidel's variables are invariant in the Gaussian approximation, (11.45)–(11.45) can also be written as

$$p = \frac{Nl_e}{z_e}X'_e - \frac{1}{l_e}X, \qquad q = \frac{Nl_e}{z_e}Y'_e - \frac{1}{l_e}Y, \qquad (11.92)$$

$$p' = \frac{N'l'_e}{z'_e}X'_e - \frac{1}{l'_e}X', \qquad q' = \frac{N'l'_e}{z'_e}Y'_e - \frac{1}{l'_e}Y'. \qquad (11.93)$$

Further modifications are necessary. If M and M_e are the lateral magnifications related to the object–image planes and the entrance–exit pupils, respectively, we can easily verify that

$$\frac{l'}{l} = \frac{R-z'}{R-z}, \qquad \frac{l'_e}{l_e} = \frac{R-z'_e}{R-z_e}.$$

Taking into account (11.90), these relations can be written as

$$\frac{l'}{l} = \frac{Nz'}{N'z}, \qquad \frac{l'_e}{l_e} = \frac{Nz'_e}{N'z_e}. \qquad (11.94)$$

When the notations

$$h = \frac{l_e s}{a} = \frac{l'_e z'}{a'}, \qquad h_e = \frac{w_e}{Nl_e} = \frac{w'_e}{l'_e} \qquad (11.95)$$

are introduced and (11.94) and (11.44) are used, (11.92) and (11.93) can be transformed as follows:

$$p = N\left(\frac{h}{z}X'_e - \frac{h_e}{w_e}X\right), \quad q = N\left(\frac{h}{z}Y'_e - \frac{h_e}{w_e}Y\right),$$
$$p' = N'\left(\frac{h}{z'}X'_e - \frac{h_e}{w'_e}X\right), \quad q = N\left(\frac{h}{z'}Y'_e - \frac{h_e}{w'_e}Y\right). \quad (11.96)$$

Using the above results and notation, simple but tedious calculations enable us to prove that

$$(p-p')^2 + (q-q')^2 = \left(\frac{N'-N}{\mathcal{R}}\right)^2 (h_e^2\mathcal{R}^2 + h^2\mathcal{R}'^2_e - 2hh_ek^2), \quad (11.97)$$

$$\frac{N}{(N'-N)^2}[(p-p')^2 + (q-q')^2] - \frac{z}{N}(p^2+q^2)$$
$$= h_e^2\mathcal{R}^2\left[L - (K-L)\frac{z}{w_e}\right] + h^2\mathcal{R}'^2_e K - 2hh_ek^2L, \quad (11.98)$$

$$\frac{N'}{(N'-N)^2}[(p-p')^2 + (q-q')^2] - \frac{z'}{N'}(p'^2+q'^2)$$
$$= h_e^2\mathcal{R}^2\left[L - (K-L)\frac{z'}{w'_e}\right] + h^2\mathcal{R}'^2_e K - 2hh_ek^2L, \quad (11.99)$$

where $\mathcal{R}, \mathcal{R}'_e$ and k are defined in (11.56).

Introducing these expressions into (11.91) and taking into account (11.65), we finally obtain the fourth-order terms of Schwarzschild's eikonal. Moreover, by applying the theorem for the addition of Seidel's coefficients (see Section 11.5), we can write

$$S^{(4)} = -\frac{1}{4}A\mathcal{R}^4 - \frac{1}{4}B\mathcal{R}'^4_e - Ck^4 - \frac{1}{2}D\mathcal{R}^2\mathcal{R}^2_e + E\mathcal{R}^2k^2 + F\mathcal{R}^2_ek^2, \quad (11.100)$$

where

$$A = \frac{1}{2}\sum_i h^4_{ei}\left[(N_i - N_{i-1})\frac{K_i}{R^3_i} + L^2_i\left(\frac{1}{N_iz'_i} - \frac{1}{N_{i-1}z_i}\right)\right.$$
$$-2L_i(Q_i - L_i)\left(\frac{1}{N_iw'_{ei}} - \frac{1}{N_{i-1}w_{ei}}\right) \quad (11.101)$$
$$\left.+(Q_i - L_i)^2\left(\frac{z'_i}{N_iw'^2_{ei}} - \frac{z_i}{N_{i-1}w_{ei}}\right)\right],$$

$$B = \frac{1}{2}\sum_i h^4_i\left[(N_i - N_{i-1})\frac{K_i}{R^3_i} + Q^2_i\left(\frac{1}{N_iz'_i} - \frac{1}{N_{i-1}z_i}\right)\right], \quad (11.102)$$

11.7. Aberrations of Axially Symmetric Systems

$$C = \frac{1}{2}\sum_i h_i^2 h_{ei}^2 \left[(N_i - N_{i-1})\frac{K_i}{R_i^3} + L_i^2\left(\frac{1}{N_i z_i'} - \frac{1}{N_{i-1} z_i}\right)\right], \quad (11.103)$$

$$D = \frac{1}{2}\sum_i h_i^2 h_{ei}^2 \left[(N_i - N_{i-1})\frac{K_i}{R_i^3} + Q_i L_i\left(\frac{1}{N_i z_i'} - \frac{1}{N_{i-1} z_i}\right)\right. \\
\left. - Q_i(Q_i - L_i)\left(\frac{1}{N_i w_{ei}'} - \frac{1}{N_{i-1} w_{ei}}\right)\right], \quad (11.104)$$

$$E = \frac{1}{2}\sum_i h_i h_{ei}^3 \left[(N_i - N_{i-1})\frac{K_i}{R_i^3} + L_i^2\left(\frac{1}{N_i z_i'} - \frac{1}{N_{i-1} z_i}\right)\right. \\
\left. - L_i(Q_i - L_i)\left(\frac{1}{N_i w_{ei}'} - \frac{1}{N_{i-1} w_{ei}}\right)\right], \quad (11.105)$$

$$F = \frac{1}{2}\sum_i h_i^3 h_{ei} \left[(N_i - N_{i-1})\frac{K_i}{R_i^3} + Q_i L_i\left(\frac{1}{N_i z_i'} - \frac{1}{N_{i-1} z_i}\right)\right]. \quad (11.106)$$

Finally, we would like to apply formulae (11.54) to evaluate the primary aberrations of an axially symmetric system. To this end, it is sufficient to rewrite (11.100)–(11.106) in terms of the ordinary coordinate y of the object point in the meridional plane and the coordinates x_e, y_e in the entrance pupil.

References

[1] A. E. Conrady, *Applied Optics and Optical Design*, Part I, Dover, New York, 1957.

[2] A. E. Conrady, *Applied Optics and Optical Design*, Part II, Dover, New York, 1960.

[3] R. K. Luneburg, *Mathematical Theory of Optics*, University of California Press, Berkeley, 1964.

[4] M. Herzberger, *Modern Geometrical Optics*, Interscience, London, 1958.

[5] H. A. Buchdahl, *Optical Aberration Coefficients*, Dover, New York, 1968.

[6] H. A. Buchdahl, *An Introduction to Hamiltonian Optics*, Cambridge University Press, Cambridge, 1970.

[7] W. Welford, *Aberrations of the Symmetric Optical Systems*, Academic, New York, 1974.

[8] H. Cretien, *Calcul des Combinaisons Optique*, Masson, Paris, 1980.

[9] M. Born, E. Wolf, *Principles of Optics*, Pergamon, Oxford, 1980.

[10] H. Rutten, M. van Venrooij, *Telescope Optics: A Comprehensive Manual for Amateur Astronomers*, Willmann-Bell, Richmond, 1999.

[11] R. Kingslake, *Lens Design Fundamentals*, Academic, New York, 1978.

[12] W. J. Smith, *Modern Optical Engineering*, McGraw-Hill, New York, 1966.

[13] D. J. Schroeder, *Astronomical Optics*, Academic, New York, 1987.

[14] R. R. Schannon, *The Art and Science of Optical Design*, Cambridge University Press, Cambridge, 1997.

[15] M. J. Kidger, *Intermediate Optical Design*, SPIE, Bellingham, 2004.

[16] V. Sacek, *Notes on Amateur Telescope Optics*, http://www.telescope-optics.net, 2006.

[17] W. R. Hamilton, *The Mathematical Papers of Sir Willams Rowan Hamilton, Vol. I: Geometrical Optics*, Cambridge University Press, Cambridge, 1931.

[18] L. Seidel, Zur dioptrik: über die entwiecklung der glieder dritter ordnung, *Astron. Nachr.*, 43, 289, 1856.

[19] K. Schwarzschild, Untersuchungen zur geometrischen optik, *Abhandl. Ges. Wiss. Göttingen*, 4, 8, 1905.

[20] H. H. Hopkins, *Wave Theory of Aberrations*, Oxford University Press, New York, 1950.

[21] M. Gaj, Fifth-order field aberration coefficients for an optical surface of rotational symmetry, *Appl. Opt.*, 10, 1642, 1971.

[22] B. Tatian, Aberration balancing in rotationally symmetric lenses, *J. Opt. Soc. Am.*, 64, 1083, 1974.

[23] G. Hopkins, Proximate ray tracing and optical aberration coefficients, *J. Opt. Soc. Am.*, 66, 405, 1976.

[24] A. Walter, Eikonal theory and computer algebra, *J. Opt. Soc. Am. A*, 13, 523, 1996.

[25] A. Marasco, A. Romano, Third-order aberrations via Fermat's principle, *Nuovo Cimento B*, 121, 91, 2006.

[26] A. Marasco, A. Romano, Houghton's camera and telescope, *Int. J. Eng. Sci.*, 44, 959, 2006.

[27] A. Marasco, A. Romano, Maksutov's cameras and telescopes, *Int. J. Eng. Sci.*, 45, 34, 2007.

[28] D. Hawkins, E. H. Linfoot, An improved type of Schmidt camera, *M.N.R.A.S.*, 105, 334, 1945.

[29] E. H. Linfoot, The Schmidt–Cassegrain systems and their applications to astronomical photography, *M.N.R.A.S.*, 104, 48, 1944.

[30] A. S. DeVany, A Schmidt–Cassegrain optical design with a flat field, *Sky and Telescope*, May, 318, 1965.

[31] A. S. DeVany, A Schmidt–Cassegrain optical design with a flat field, *Sky and Telescope*, June, 380, 1965.

[32] R. D. Sigler, Family of compact Schmidt–Cassegrain telescope designs, *Appl. Opt.*, 13, 1765, 1974.

[33] R. D. Sigler, Compound Schmidt–Cassegrain telescope designs with nonzero Petzval curvatures, *Appl. Opt.*, 14, 2302, 1975.

[34] R. Gelles, A new family of flat-field cameras, *Appl. Opt.*, 2, 1081, 1963.

[35] R. D. Sigler, A new wide-field all spherical telescope, *Telescope Making*, 30, 4, 1987.

[36] V. Mahajan, Zernike annular polynomials for imaging systems with annular pupils, *J. Opt. Soc. Am.*, 71, 75, 1981.

[37] V. Mahajan, Strehl ratio for primary aberrations: some analytical results for circular and annular pupils, *J. Opt. Soc. Am.*, 72, 1258, 1982.

Index

Abbe's condition, 79
Abbe's invariant, 27
aberration function, 49, 51
aberration vector, 52, 201
aberration-free system, 6
aberrations, 201
achromatic, 10
achromatic doublet, 145, 149
Airy disk, 9, 87
Airy ring, 88
angular function, 195
anterior base plane, 195
anterior focal length, 33
aperture, 101
aperture stop, 34
aplanatic, 84
apochromatic, 10
apochromatic doublet, 149
astigmatism, 71
Astigmatism and curvature
   of field, 69
auxiliary optical path, 49
axial chromatic
   aberration, 9, 77

back focal, 33
bending, 146
Buchroeder camera, 168

Cassegrain combination, 97
Cassegrain telescope, 94
characteristic functions, 49

chromatic magnification
   or lateral color
   aberration, 77
Coma, 68
conic, 13
conic parameter, 14
conjugate planes, 6, 27
curvature of field, 73

Dall–Kirkam combination, 97
diffraction limited, 9
directrix, 14
Distortion, 72
distortion, pin-cushion and
   barrel, 72

eccentricity, 14
eikonal equation, 185
encircled energy fraction, 88
entrance pupil, 6, 34
Euler–Lagrange
   equation, 178
exit pupil, 6, 35

Fermat's principle, 2, 171
fifth-order aberrations, 202
flat-field camera, 159
focal surfaces, 71
focus, 14
Fraunhofer doublet, 145
front focal, 33
full-aperture corrector, 162

223

Gaussian approximation, 8, 26
Gaussian image plane, 8
Geometric optics, 1

Helmholtz condition, 79
Helmholtz's invariant, 37
Houghton's combination, 110
Huygens' principle, 189

ideal system, 6
image space, 7

Klevtsov telescope, 164

Lagrange's invariant, 38
Lagrange's optical invariant, 184, 200
longitudinal spherical aberration
    curve, 13

magnification, 7
magnification factor, 28
Maupertuis' principle, 182
meridional plane, 6
meridional ray, 7
mixed functions, 196
monochromatic aberration
    functions, 11

nodal points, 33

object plane, 5
object space, 7, 193
optical axis, 197
optical path length, 2, 171
overcorrected, 9

parabola, 15
paraxial approximation, 8, 26
Petzval's curvature, 73
Petzval's theorem, 73
point principal function, 193
point spread function, 87
posterior base plane, 195
posterior focal length, 29, 33
power, 28
Pressman–Camichel combination, 97

primary aberrations, 54
primary axial color, 9
primary lateral color, 10
principal functions, 192
principal planes, 33
principal ray, 35

ray, 2
ray tracing, 11
ray-intercept curve, 12
reduced coordinates, 206
reflection law, 180
refraction law, 179
reversible, 198
Ritchey–Cretien combination, 97

sagittal plane, 6
sagittal ray, 7
Schmidt camera, 103
Schwarzschild's eikonal, 49, 208
secondary spectrum, 10
Seidel variables, 207
semiapochromatic doublet, 149
sine condition, 79
skew ray, 7
speed, 102
spherical aberration, 67
spot diagram, 13
Steinheil doublet, 145
stigmatic path, 54
stigmatic points, 4, 6
surface of revolution, 13

tangential plane, 6
tangential ray, 7
third-order aberrations, 54, 202
triplet, 152

undercorrected, 9

vignetting, 35

wavefront, 2
Weierstrass–Erdman jump
    conditions, 178
Wright camera, 109